Worm Bins

The Experts' Guide To Upcycling Your Food Scraps & Revitalizing Your Garden | Worm Composting & Vermiculture Made Easy

Your Backyard Dream - Book 1

By Geoff Evans

Disclaimer Notice:

Please note the information contained within this document is for educational and entertainment purposes only. All effort has been executed to present accurate, up to date, and reliable, complete information. No warranties of any kind are declared or implied. Readers acknowledge that the author is not engaging in the rendering of legal, financial, medical or professional advice. The content within this book has been derived from various sources. Please consult a licensed professional before attempting any techniques outlined in this book.

By reading this document, the reader agrees that under no circumstances is the author responsible for any losses, direct or indirect, which are incurred as a result of the use of the information contained within this document, including, but not limited to, — errors, omissions, or inaccuracies.

Table of Contents

Your Free Printable Worm Bin Label!

As a thank you for downloading this eBook, I'd like to give you this printable worm bin label. It's super helpful to stick on the front of your bins for when your family members (who might not have read this book) go to put something in the compost bin – they'll get a quick visual reminder as to what is allowed in and what kinds of waste they shouldn't put in your bins.

You can download the printable label here - https://bit.ly/wormbinfreegift

No email sign-up required. Download & Print straight from the Google Drive link.

With that, let's get to the wiggly stuff!

Introduction

You've bought your worms. You've got stacks of old newspapers and cardboard boxes ready. You have a five-gallon bucket or two, with holes questionably drilled into them. You've been saving all your coffee grounds and banana peels for the last month. You're ready to start worm composting. Or, are you? The bin is stinky, there's so much liquid coming out of it, and it's getting really warm, all of your worms are trying to escape and you're not even sure what this "casting" stuff should look like. Or perhaps you haven't even gotten that far; maybe you've just read a few articles about worm composting (also commonly referred to as vermicomposting), with no clue how you should get started. Are worm castings compost? Or manure? And why does it have so many different names? Vermicast, Vermicompost, Worm Castings, Worm Compost. It's easy to get overwhelmed! It can feel a little bit like staring at the pasta aisle in a fancy grocery store: you don't even know where to start. On the other hand, you know how beneficial this practice is; it reduces your carbon footprint, it reduces consumer waste, it helps your plants and garden, and you could even sell your worm castings to make some extra cash! It seems like the perfect hobby, and you're eager to pick it up. But how?

As a beginner worm composter, you may have all of these questions, and more. Luckily, all of the answers lie in the following pages. This book is the definitive guide to vermicomposting. You'll learn how to start your bin, how to maintain it, what to use inside it, what different issues might indicate and how to go about fixing them. You'll also learn some of the science behind what makes a worm bin churn. Even if this is the first time you've ever heard the phrase "worm composting," by the end of this book you'll know the process inside and out – so that you could do it with your eyes closed.

I've always been passionate about the environment. I've been composting in the conventional way for years. My housemates used to hate me for it, but I love being able to turn what normally would be trash into something that could breed new life. When I discovered vermicomposting, I was hooked. It's become more than a hobby to me; it's a lifestyle. Worm composting fuels my gardening and my gardening in turn fuels my healthy eating habits. My whole lifestyle is sustainable, renewable, and beneficial for my family. All because of a few little worms! Now, I get to share this love with you. If you are anything like me, and want to do your part for the environment, then learning how to compost with worms will change your life for the better. You may come to be known as the "worm guy" or the "worm girl" by your friends, but trust me - it's well worth it. Those same friends definitely won't be complaining when you make dinner with your homegrown, healthy, and nutritious veggies, or when you give them some of your castings to help fuel their gardens.

Worm composting has a whole host of benefits. It reduces the waste produced in the home, it revitalizes soil in the garden, and its small-scale nature generates significantly lower methane emissions than industrial-scale composting. Furthermore, it doesn't require the use of any toxic chemicals to speed up the decomposition process. Worm composting is an incredibly simple way to help save the environment, and add 'rocket fuel' to your soil. If you are concerned about your carbon footprint, by the amount of waste still going to landfill, or the sustainability movement as a whole - then vermicompost is the perfect place to start.

Reading this book will take you from novice to expert vermiculturist, in no time. Here, you will learn the basics of worm composting, as well as some advanced techniques that I've picked up over the years. You'll learn how to sustain a healthy culture, the fundamental do's and don'ts, common misconceptions, and various unique applications. You'll discover what microbes are and how they steer this whole ship. You'll find out how to teach vermicomposting to others, and how to make it a fun activity to do with the entire family. It may seem like a lot to learn from one book, but don't worry. The information is broken down and explained in a clear and approachable way. While the book assumes the reader has zero to little experience with worm composting, those with higher levels of expertise can still *dig up* (pun intended) some things they didn't yet know!

You should read the book in its entirety before starting your worm composting journey. A good understanding of the basics will set you apart from those who learn as they go. While there will still be plenty of trial and error, as is necessary when learning anything new, it will be much easier to figure out what went wrong and how to fix it - with the knowledge you gain here. Keep the book somewhere handy for the first few months of your composting journey, too. Just like studying for a class, refreshing your memory will make learning easier, faster, and much more comprehensive. Remember: it's okay to make mistakes. It's okay if your compost is stinky, if you forget to feed it, if it overheats, if it's too moist. The most important thing is to learn. Mistakes are a good thing if you know how to correct them.

As you begin reading, keep in mind the KEY goal that all composters share: to create a more sustainable household. Remember this goal during the frustrating or confusing times. Composting, with the right mindset, can be a selfless and even noble pursuit. It may not be for everyone, but it sure does help us all. Read this book, follow the guidance, and know that you are not only helping yourself and your family, you're helping precious Mother Earth.

Chapter 1 - What is Worm Composting?

When you stroll through a beautiful green garden, or glance at a pristinely kept lawn, you probably don't think about the brown underneath all that green. But lying just below the surface is a world full of life that makes all that beautiful landscaping possible: the earthy soil - feeding lawns, gardens, and landscapes all across the world. Driving all of that growth is arguably one of the most important species on the planet: earthworms.

Vermicomposting first began in the early 1990s. Originally, its purpose was to clean agricultural and industrial waste before it entered large bodies of water. As research into worm composting continued, it was revealed that the castings which worms left behind were incredibly rich in nutrients. They could be used not just to prevent ecosystem deaths, but to give struggling plants and soils new life. The byproduct of worm composting turned out to be a superfood for the earth! The first vermicomposting system in the United States was located in Portland, Oregon. It still operates to this day. The plant uses a continuous flow system (which you will learn about later) to compost over two thousand tons of food per year.

Earthworms are a keystone species. A keystone species is one that acts as the foundation of an ecosystem. If it were removed, the ecosystem would be drastically altered, and would risk failing entirely. Earthworms perform several ecosystem 'services' that make them vital to the survival of a healthy system. When they burrow, they improve the *shape* of the soil itself. Their burrows create air pockets, 'fluffing' the soil, which allows more water to fill the gaps, hence improving drainage. Soil that has been aerated by earthworms can <u>absorb as much as ten times more water than non-aerated soil</u>. This increased water content brings more water and nutrients to the plants growing in the soil.

Earthworms also serve as decomposers, they break down large organic materials into their original chemical components, when they eat and later pass them. While decomposition is a natural process, earthworms act as a catalyst to speed up the process. This composting superpower is the primary goal in vemiculture. By feeding your worms food waste and other decomposable matter, they turn it into nutrient-rich "castings," or excrement, that has the potential to increase soil yield by as much as 30%. It's one of the most potent fertilizers in the world!

When we think of animals that are emblematic of sustainability and environmentalism, we often think of endangered species and our efforts to conserve them. Take Pandas and Polar bears for example. It's easy to understand why. They look fluffy and cuddly. The image of a skinny polar bear on a small piece of ice is emotionally striking. But earthworms are incredibly more ecologically important to the environment. Industrial farming, pesticides, and mono-cultures all contribute to the degradation of topsoil in most industrial farms. This soil is left with only trace amounts of nutrients. It can't hold water as effectively as it should. And the crops that grow in it, suffer the consequences. Preserving earthworms in these areas, introducing them back to the ecosystem, and farming their glorious castings is an excellent way to address and solve these soil issues.

Worm composting is simply the process of feeding food waste to worms in order to collect their castings. Unlike traditional composting, vermicomposting uses worms as a catalyst to speed up decomposition. The worms also enhance the resulting compost, as they create nutrient-rich humus - the nutrient-filled organic part of soil. As the food waste travels through the worms digestive systems, microbes break it down into its base components, carbon and nitrogen. When the worms pass their food, they leave behind castings. Getting your worm bin right takes a small financial investment, some understanding, and a little time dedicated to monitoring, but if done right - as this book explains - worm composting is an incredibly rewarding process! A typical vermicomposting setup consists of a storage device, a way to drain the storage device, worm bedding, food waste, and, of course, worms. Each of these components will be discussed in further detail in Chapter 3.

As just briefly mentioned above, another key player in composting, and indeed any form of decomposition, is the microbes. Microbes are microscopic organisms consisting of bacteria, viruses, or fungi. Most crucial for composting, are bacteria, which help to decompose food waste. Bacteria break down food into its base organic components. Furthermore, the bacteria act as a filter to detoxify the worm waste, ensuring that the compost is beneficial to the plants that use it. Since worms don't have stomachs or teeth, they rely on their relationship with bacteria to break down and process their food. Because of this, bacteria are the best friends of any home composter. If your worm bin is in good health, it will have its very own community of microbes. Worms aren't the only organisms that use bacteria to break down food, many animals do, including humans. Studies of bacteria in the human digestive system are still early in development, but researchers have already linked certain gut bacteria to a wide range of health risks, including depression, obesity, and some forms of cancer. Healthy (and unhealthy) decomposition uses a plethora of bacteria, no matter whose gut it happens in!

If you happened to have done some of your own research already, then you might have come across the concept of "worm tea." It is not, as the name may suggest, something you should drink. It is, however, an excellent source of nutrients for your plants. Worm tea is a liquid made by steeping some of your castings in water, then mixing it with a slow-release source of sugar, such as kelp. As it steeps, the microbes bloom into a flourishing colony that, when used to water plants, powerfully fertilizes them. Plants absorb the nutrients from the vermitea (worm tea) faster than from soil, since it is sprayed directly onto the leaves or onto the base of the plant. It can also act as an organic pesticide. It repels sucking and chewing bugs such as aphids and whiteflies from the leaves of plants. A plant regularly squirted with worm tea is a healthier one. Just like a person taking vitamins, this plant is now better equipped to handle harmful attacks. It also encourages the cuticle of the leaf (the waxy, outer layer) to grow, giving the plant further defense against pesky bugs. The cuticle also serves to reduce moisture loss. Making worm tea is the same process as making cold tea. Simply put some castings in a porous bag, such as cheesecloth or an old shirt, set it in some water, and steep for at least 24 hours. You could instead skip the bag (like I do), but just make sure to strain the tea before using it. You can then put the liquid tea in a sprayer or watering can. It doesn't store very well, as the bacteria have a short lifespan, so plan ahead and try to use it within a day or so.

All that said, why would you go to the effort of raising worms if you could just compost without them? Why would you use vermicomposting over regular composting? Let's take a look at some of the comparisons. While vermicomposting focuses on the product of worms, regular composting is solely driven by microbes. Traditional compost piles, as they decompose, heat up, cure, and then cool. These piles can reach very high temperatures, and some industrial outdoor composting sites even have to monitor their piles for spontaneous combustion. The piles must be churned manually, adding to labor requirements. Since heat is part of the process, you need quite a lot of material in order to form a thick enough layer that the pile will heat up enough in the first place. This means that traditional composting is more labor intensive, resource intensive, and potentially more risky than vermicomposting. When normal compost heats, it also releases methane gas (CH_4). As a very potent greenhouse gas, methane is harmful to our environment in large quantities. If you're trying to reduce your carbon footprint, that should be a very big deterrent. Because the worms are constantly moving around and eating waste, food decomposes faster in vermicompost, and methane doesn't get as much of a chance to develop. Regular composting also takes significantly longer than vermicomposting. The worms do so much of the work, that they can do in eight weeks what normal composting takes six months to do! On the other hand, you can't feed the worms all of your waste; they have a broadly specific diet, which we'll talk about later. The benefits of worm composting far outweigh the disadvantages. Vermicompost yields better and faster results, is more rewarding, and can be more profitable than traditional composting. <u>Did you know that a cubic yard of vermicast can be sold for as much as $400?</u>

Hopefully, you now have a basic understanding of what vermicomposting is and what it entails. What follows in this book will continue to explain all of these details and more, as well as uncovering some top secrets I've learned over the years.

Chapter 2 - The Benefits of Worm Composting & Common Concerns Beginners Have

In college, I composted religiously. There was always a bucket next to the kitchen window in my various apartments, filled with smelly food waste that I insisted my resentful roommates contribute to. I didn't manage it very well. The bucket, a simple flowerpot, wasn't even big enough to let the compost heat up. I didn't know anything about curing or aerating. I just poured my leftover pasta into the bucket and let it ferment. I would then proceed (since I was in an apartment and didn't have access to a garden) to walk through campus and sprinkle the wet mass into various flower beds. I'm sure all that did was get the squirrels *drunk*, but at the time I thought I was a master gardener! When I got older, I didn't need to have roommates anymore.

Without people around to complain, my compost addiction got even more prevalent, with large buckets taking up most of my kitchen. I got used to the smell, but when I brought home significant others, or even my parents, they didn't usually stay for long. They were always less than impressed by the stinky buckets all over my apartment. My compost was causing issues, so I decided to do some research.

At first, I was just looking for ways to get rid of the smell. What I stumbled across though, was something far better. Worm composting. Reading blog post after blog post, it grew clear to me that this is what I should be doing, rather than the method I had become accustomed to. I went all in. I ordered worms from the local pet store, I took a trip to the hardware store to get 5-gallon buckets and an electric drill to put holes in them. I took extra cardboard and newspaper from every friend who had some to spare. I even froze my food waste, so I could guarantee a steady supply of food to my little worm friends. That first haul, after the slowest few months of my life, was so exciting. I practically ran as I carried my bucket to the balcony. Sifting through the worms to pull out the castings, scooping them out, and sprinkling them over the small flower trough I had, you'd think I had just received a Medal of Honor with how proud I was. My family and friends were happy too, although I think that's mostly because I got rid of that awful smell.

I'm writing this book because I want you to experience that same sense of pride and self-satisfaction. Building something with your own two hands is incredibly rewarding. Seeing your garden flourish is beautiful. For me, even a few years later, I am filled with pride just looking out in my garden and seeing those bright colors, thick stems, and happy plants. My lawn is the envy of the neighborhood. I even make a few extra bucks with my leftovers, so my wallet is happier too. I hope that, just as your vermicompost feeds your garden, so it will feed your soul.

The benefits of vermicompost are plenteous. Some benefits, like the fertilization of plants, are obvious. Others, like pride and self-satisfaction, or admiration from your neighbors, are less so. With that, let's discuss the plethora of benefits of worm composting.

Worm composting can (but isn't limited to) be done successfully with minimal equipment. A few buckets, paper scraps, and some worms are enough for you to start. Additionally, since vermicompost does not require heat cycles, it can be kept anywhere. A garage, a basement, a kitchen, or a balcony - as long as temperatures are comfortable for you, your worms will likely be comfortable, too. If you're feeding your worms properly, your compost won't be odorous, unlike normal compost piles. Worm composting is also very child friendly! They can learn how to care for animals, how to responsibly maintain the compost, and even how gardening works. Compare that to traditional composting, where the kids would only learn how to put their leftovers in a bucket and stir it. Ultimately, the largest benefactor of vermicompost is you. You, in undergoing this endeavor, will learn a new appreciation for nature. You'll develop a new understanding of ecosystem relationships. You'll gain experience growing bigger, healthier crops than you ever have before. You'll earn the pride and satisfaction that come with creating something good for the earth, with your own two hands.

Worm composting isn't all sunshine and rainbows. It's also sifting through piles of worms to pick out their waste. There are, of course, some disadvantages to worm composting. First is the effort required. Since you are dealing with living animals, vermicompost requires more attention and care than traditional composting. Also, since the temperature worms enjoy are the same ones we are comfortable in, the compost needs to be monitored to ensure that it doesn't overheat in the summer, or get too cold in the winter months. It can also become more expensive than traditional composting. Purchasing worms is the obvious one, but there are other tools you can buy to enhance your worm bin and to carefully monitor factors such as moisture, pH, and pests - which all come with a price tag. As your vermicompost endeavor grows, you may feel the need to upgrade existing equipment, purchase new tools, or even expand the scale of your compost.

It's common to hesitate before committing to any new project. Never mind one involving worms! Many new worm composters are concerned about smells. My family was, too. Fear not. Unwanted odors can be avoided with proper feeding. Worms are hungry critters. They eat pretty quickly. But if your compost is stinky, that's a sign that you are feeding the worms too much, and the food is rotting before they get the chance to eat it. As you'll learn in later chapters, it's better to underfeed your worms instead of overfeeding them, especially with a new bin. If you get odors, give your worms less food. If you have food scraps, but your worms haven't finished what you've already given them, freeze the food to use at a later date. This way, you can space out your feedings to keep your worms eating at a steady pace. Each worm bin is unique, so you'll have to do some experimenting to find out what the correct amount of food is. A good rule of thumb is to start with a quarter of the weight of your worms in food. As your bin grows, it will process waste faster, and you may find yourself needing to add more than this; but at first, use the quarter weight rule and adjust as needed. If you do, odors won't be a concern.

Another common concern is that the worms will escape the compost and get into your house. This is a valid concern. Nobody (even me) wants a clew of worms crawling around their house. Luckily, while a beginner vermiculturist may be concerned about this, it's not as big of an issue as it may seem. The best way to prevent worms from escaping is to simply have a lid on your bin. You can use a porous cloth, or use a solid lid with holes cut in it. The worms won't be able to escape, and if you're feeding them properly, they probably won't want to! Worms are averse to light, so even a lamp above your bin will discourage escapes. Why would they want to get out if they have a nice, dark, comfortable bed filled with food that they can burrow in? If you have worms gathering at the top of your bin, that can be a sign that the ecosystem is off balance. Your bin could be too moist, too alkaline, have too much food, or could simply be overcrowded. As you read this book, you'll learn precisely how to address each of these issues, as well as several more.

If you're anything like me, just knowing this much information has you itching to get to the hardware store. What should you add to your list? What should you buy? Should you order that $300 worm bin from Amazon? In the next chapter, you will learn about the essential and helpful equipment to make sure your worm bins are a success.

Chapter 3 – What Equipment Will You Need?

Worm bins can range from a simple DIY box to a high-tech, mass-production farm. What are the tools, equipment, and supplies you need for your compost to succeed, regardless of scale? If you follow these guidelines, you'll find it's easy to get your first worm bin up and running.

The first thing you need is something to hold your vermicompost. This can be a 5-gallon bucket, a storage box, or even a specialized worm compost container. The most cost-effective option is a simple plastic box. A wide box works best since it has a larger surface area to spread the food waste over. That means you can feed your worms more efficiently. Shallow, wide boxes also help prevent overheating. We will talk about how to choose the right size and type of box for you, in Chapter 4.

The next thing you will need is paper scraps. Newspaper and cardboard both work well. Then, you need a way to elevate the box so that it can drain. You can easily use something like a block of wood or a couple of bricks for this. Next, you need a tool to put holes in the box. You'll need holes in both the lid and on the bottom, so a drill clearly works best here!

Worms breathe through their skin, so it's very important that the bin has ventilation and that the compost isn't overly wet. Holes in the lid of the bin, or a porous cloth cover, will let air into the bin. Holes in the bottom will aid leachate drainage. Leachate is the excess moisture produced in the vermicomposting process. It is undesirable, but inevitable. Elevating the bin and putting holes in the bottom lets the leachate drain out of the bin. Unless you store your bin outdoors, you will want to invest in a tray to place under the bin to catch the leachate as it drains out. Alternatively, a second lid works well.

Remember that **leachate is not worm tea**; it is a byproduct of the production of castings. When you first build your compost bin, please ensure sure you have built a draining system. It's a mistake that a lot of beginners make, no matter how silly it might seem!

If making your own bin sounds a little too time-consuming or difficult, you can buy a pre-made worm composter. Some of these are stackable trays, and some are cloth netting, or a bucket with built-in holes and a *spigot* for draining. These can run from as low as $45 to upwards of $300. If cost isn't an issue, this is a great way to start your vermicomposting venture. If you go via this route, all you would need then to purchase are worms and supplemental supplies for regulating important aspects of the bin's ecosystem. Pre-made bins often include trays to separate the castings into different layers. When conditions are right, the worms eat their way to the top of the bin, and the castings on the bottom tray don't need to be sifted through. If you're interested in cutting down some labor and don't mind the price, then a pre-built bin is best for you. If price is more of a concern, building your own bin is still incredibly effective. And it's rewarding.

There is plenty of supplemental equipment you can invest in later on. This ranges from sensors to monitor conditions, to sifters, to labor-savers. Sensors are a huge help. A thermometer lets you measure the compost's temperature so you can ensure the worms are comfortable and take relevant action if not. I'd say that a moisture meter is essential. Overly moist, or dry bins, are both detrimental to the worm's environment. You'll learn more about this in future chapters. A pH monitor will also come in useful. A small shovel to move worms and food waste around is good to have on hand. 'Mix-ins' are available to purchase too. Lime mix, for example, can help if your worms are being fed lots of acidic food. Make sure that the mix you get has a calcium carbonate (CaCO3) concentration of at least 95%. Food grade 'diatomaceous earth,' which is made from microscopic ocean animals called diatoms, is both a good source of pH-balancing calcium and prevents hard-shelled pests from bothering your worms. The small particles of calcium act like sandpaper to the shells of insects such as pill bugs or beetles. You can also add a compost accelerator like 'bokashi bran' or a slow-release sugar like kelp. These will boost the speed at which organic waste breaks down in your bin. Depending on the amount of waste your household produces, and on the concentration of worms in your bin, this may or may not be helpful. If you have a stinky bin and your freezer is full of scraps, you should almost certainly invest in an accelerant!

You may also need a sifter. When you're ready to harvest, you need to sift the worms from the castings, as you don't want to accidentally lose any of those beautiful little black-gold makers to your garden. To do this, you can sift the pile by hand, using a light to repel the worms from the sides of the pile. Alternatively, get yourself a sifter – it's what I have. There are several pre-made ones available for purchase, but just like the pre-made bins, they are expensive. The cheapest way to sift is to use a sieve, like those found in the cooking aisle of the grocery store. If the sieve you find is too fine, you can cut small slits in it. Just make sure that the holes you make aren't big enough for the worms to get through. You'll learn more about how to harvest your castings in Chapter 8, so stay tuned for that.

Once you have all of your materials, it's time to put it all together. You'll start by drilling holes into the top (the lid) and bottom of your bin. The holes don't need to be bigger than a standard drill-bit. Space the holes out in a grid 2-3 inches apart. You can make the holes the same on the lid and bottom of the bin. Then, you need to find a good home for the bin. Put it somewhere stable and prop it up on bricks or wood so the leachate can drain out into a spare lid or container. Leachate will drain on its own, without any encouragement on your part. A normal amount of leachate is completely different from bin to bin, so when you first start your bin, monitor the amount of leachate carefully. If it changes without you increasing the worms' amount of food, you may need to address the moisture levels. Ensure your bin is out of direct sunlight. The temperature should be around 40-80F (4.4-26C). Finally, before you add the worms, you need to add your bedding.

It's better to have your bin built and ready before you go and buy your worms. If you get your worms first, you'd need to find a place to store them while you get your bin prepared. Building your bin ahead of time means that all you have to do once you buy your worms is simply add them in! When you add your worms, be sure that there is also a little bit of food in the bin. After a few days, your worms will begin happily eating your food waste.

Chapter 4 – Designing Your Worm Bin

Now that you have all of your key materials purchased, collected, and set up, how do you make your worm composting *station* successful? What measurements or physical conditions must you keep in mind? Can you make a successful worm bin with just a box and some worms? It's normal to see all your supplies gathered and feel a little aimless. Don't worry, not only will it make sense soon, it will seem almost intuitive. Let's discuss how to build your vermicompost bin and keep it running. At the end of the chapter, there's also a very unique DIY worm bin build you can learn how to create!

First, you should be closely monitoring the physical conditions of your bin on a fairly regular basis. I recommend every 2 or 3 days if you can. We'll get onto what to do when you want to take a vacation later on, as part of the FAQ's. But for now, the three most important physical conditions to monitor are acidity, temperature, and moisture. The pH scale measures how acidic or basic something is, and runs from 0 to 14. Pure water is completely neutral, so it has a pH of 7. Lemon juice, which is very acidic, is around 2. Bleach, which is very basic (alkaline), sits at 13. The acidity of your bin should range from slightly acidic to barely alkaline. Keep your bin in the 6.5–7.5 pH range for the optimum worm environment. Compost that is too acidic or alkaline will harm your worms' skin and can even cause burns. Next, make sure you keep the temperature of your bin comfortable. Worms are comfortable at about 40 - 80 degrees Fahrenheit, as mentioned previously. Watch out for your pile becoming too deep, as dense compost layers can lead to increased temperatures. Making layers of worm bedding and using a bin with a large surface area can help you control temperatures in your bin. Temperature is also influenced by the location of your bin. If it is in your house, you may never have to worry about temperature. If it's in your garage or outdoors, it is susceptible to dangerously high or low temperatures. Take the temperature of the compost regularly. If your bin is too hot, you can use frozen water bottles tucked into the corners to help cool it down. If your bin is too cold, and moving it to a warmer place isn't an option, you can overfeed it or use a bin with a smaller surface area. This technique will not erase the aforementioned issues that overfeeding causes, but it will keep your worms from freezing. Large amounts of bedding act as insulation too. This will trap air in little pockets, bedding is excellent at regulating temperature. So regardless of whether your bin is too hot or cold, adding more bedding is a crucial step for temperature control.

Finally, it's important to carefully monitor the moisture levels in your bin every few days. Your vermicompost could become overrun with other critters that break down food before your worms get the chance to eat it. Additionally, since worms breathe through their skin, a bin that's too moist risks suffocating them. This happens if the compost becomes too dense, and the air stops diffusing properly. Ideally, your bin should have a moisture reading of around 70–80%. You can measure this with a moisture meter, available at most plant nurseries or online. Other methods of measuring moisture require complex math or estimation, both of which are inconvenient or risk inaccuracy. If your bin becomes too hot or too cold, don't try to control the temperature by adding cold or hot water, as this will disturb your moisture content. A sign that your bin is too moist is large amounts of leachate production. The definition of a large amount varies from bin to bin, but if your bin is producing more leachate than normal, that's an indicator to check the moisture level. Worm bedding or stackable trays both help regulate moisture. If you have an overly moist bin, add absorbent bedding in layers, or get a bin with trays. If your bin is dry, you can give your worms some extra food. High moisture content fruits such as watermelon have worked well for me.

Regardless of whether you decide to use a pre-made or homemade bin, there are still a couple of other considerations you need to take into account. These include the size of the bin, building material, and where to buy it from. The bigger the bin, the more worms and compost you should have. Make sure, if you decide on a big bin such as a large storage box or an old garbage can, that you have plenty of bedding layers. The bedding layers should make up half the volume of the bin. This will help prevent overheating. Once your bin has been going for over a week, you'll probably be feeding it half a pound of food per pound of worms per day. This means that, if your household produces half a pound of food waste per day, you should have a big enough bin to fit one pound (just under ½ a KG for our English folks) of worms. This space is fitting for a household of one or two people. If your household produces lots of food waste, or simply has more people in it, it will be much easier to maintain a larger bin. A smaller bin will process food slower, so if you produce less waste, small is the way to go. Sources for bins include local hardware stores, stores that sell storage containers, or online retailers such as

Amazon or niche gardening websites. You can even make a worm bin out of an old wheelie bin! A few retailers also offer bags instead of bins. The Urban Worm Bag, from the company of the same name, is a great example. This type of bin is best purchased pre-made, not homemade, since materials used can be niche and difficult to acquire without ordering in bulk quantities.

The main function of bedding is to increase the amount of carbon in your bin - relative to the amount of nitrogen. Carbon and nitrogen are both byproducts of decomposition. While the precise measurement isn't of concern, know that the more bedding you add, the better ratio you'll have. You can never have too much bedding! The bedding also serves as a food source for your worms. When you add it to your bin, add it in layers separated by food so that it isn't all bunched at the top or bottom. This will encourage the worms to consistently move up the bin as they eat, **separating the castings and food in your bin, even if you don't have trays.** The bedding will also make your compost more porous, making it easier for your worms to breathe. If you find yourself wondering if you should add more bedding, just do it! If you think you probably have enough bedding, add a little more. If you think you have too much bedding, then now you probably have the right amount.

Worm bedding is where you put to use that newspaper and cardboard we mentioned earlier. You can use other materials, too. The best beddings are those which are still compostable, but that also won't dissolve after a day in the rain. Leaves, moss, and yard clippings also fit the bill. Shredded paper is a great starting place for your bedding. Let your worms eat your bank statements! Junk mail and newspapers are nice sources of paper that you probably get for free. Make sure you don't use glossy, coated paper, or bleached office paper, as this will decompose differently and can pose a health risk to your worms. Paper is a great choice because it's versatile, and it's incredibly easy to find. Cardboard is another awesome option. Since it's thicker than paper, it will take longer to be broken down by your worms. You can find corrugated cardboard from delivery packages, mail suppliers, or your local grocery store. The main disadvantage of cardboard is that it's difficult to shred without a knife or some muscle. Dry leaves are a great alternative, especially if you live somewhere with trees that shed their leaves during the colder months. As with bringing anything from the outside in, there could be bugs and other critters living in those leaves. Be careful, as leaves aren't as moist as paper and cardboard. A dry compost pile produces less leachate than a moist one and will be uncomfortable for your worms. Make sure you regularly measure the moisture level of your bin. If it does become too dry, you can moisten some paper or cardboard and add it as bedding. Leaves and other yard clippings also tend to heat up quickly. Avoid simply dumping piles into your bin. Always spread your bedding out and distribute it evenly throughout the bin. Another common bedding material is the processed husk of coconuts, called coir. Coconut coir is readily available from stores; it holds an incredible amount of moisture and breaks down slowly. Finally, if you have a half-finished compost pile without worms, you could add it to your vermicompost as bedding. You can only do this if it's not decomposed to the point of resembling soil. At that point, the compost will be too dense for the worms to burrow through. Whichever bedding material you choose, as long as you use lots of it, your vermicompost will be much better off.

Another choice to make is whether to go large or small-scale for your vermicomposting endeavor. Should you have one bucket, or several large troughs taking up your backyard? It is entirely up to you. If you go small scale, know that your worms may outgrow the bin you have them in. In that case, you can transfer extras to your garden, add a second bin, or get your neighbors involved in vermicomposting. If you want to go large-scale, you will, of course, need greater financial investment. Larger scales require more worms, more bins, and more supplemental supplies. If you are using vermicompost in a neighborhood garden or to teach a class, bigger will be better. If you are just composting at home, start small, and if you find yourself wanting more castings, then simply start a new bin.

It may feel difficult at first, to maintain your bin. You may forget to measure moisture, you may add too much citrusy food and make your pH divebomb, you may forget to bring the bin indoors during an overnight freeze. These mistakes are normal. Don't prevent yourself from experiencing the magic of worm farming just because you might make a few mistakes. Read this book, learn, and prepare. That way, when you do make mistakes, you can identify them and grow. The more you do this, the easier it will be, and the better you will become at vermicomposting.

It's not uncommon to have old garbage cans or wheelie bins lying around your yard that you no longer use. You can turn these into composting bins, too. Drill holes in the lid the same way you would with any other container. Add holes around the upper sides of the bin. Add holes in the bottom, too. Fill the bottom 2 inches of the bin with gravel so that the holes don't clog. Make a lattice of wooden beams inside the bin. Lay boards alternating north-south and east-west in the bin. This will create air gaps. The bin will be very deep compared to usual, so this lattice is necessary in addition to filling half the bin with bedding. Layer the bin with food waste, bedding, and worms, and voila, your one-of-a-kind conversion is complete!

Chapter 5 - Getting Your Worms, But Which Breed & From Where?
(A Country by Country Guide)

Now that you have your bin picked out, your bedding material selected, and your supplemental supplies purchased, what about the most important 'ingredient' for your worm bin? The worms! How should you buy your worms? Which breed should you get? Is it okay to buy them online? If it seems peculiar to buy a box of worms from a website, that's a totally normal feeling. At the end of this chapter, you'll know what's safe for your worms, what type of worms to get in the first place, and even some websites to buy them, depending on which country you live.

When you buy worms, make sure to use their Latin names. Some species have similar nicknames but are completely different worms. Several different breeds, for example, are known *colloquially* as Red Worms. The most common (and most ideal) vermicompost worm is the Red Wiggler, also known as the *Eisenia fetida*. These worms are widely available. They can survive temperatures as low as 40F and as high as 90F. Furthermore, Red Wigglers are not invasive, so in the unlikely event that a few manage to escape your bins, you don't have to worry about your garden health. Tiger worms, or *Eisenia andreia*, are also a very common choice for worm composters like us. Tiger worms are around the same size as Red Wigglers. The main difference between these two species is cost; Tiger worms are a little more expensive than Red Wigglers. Other good options are the European and African Nightcrawlers, *Eisenia hortensis* and *Eudrilus eugeniae* respectively. These worms are typically much bigger. European Nightcrawlers excel in cold areas, they are mostly used for fishing in cold waters since their temperature tolerance is so robust. African Nightcrawlers don't share that cold temperature tolerance, but instead, they can tolerate up to 90 degrees Fahrenheit and are generally stronger than their European counterparts.

The aforementioned worms are all great choices for your vermicompost. Which one you choose is up to you. There are some species, however, that are not recommended for worm bins at all. Driftworms, or the *Lumbricus rubellus*, is not a compost worm but a burrowing one. Additionally, it is an invasive species, and may not even be legal to purchase in your area. Common earthworms, called *Lumbricus terrestris*, are also burrowing worms and won't compost your waste very well. Some vermicompost guides recommend Indian Blue worms (*Perionyx excavatus*), but while these are fast breeders and do compost, they are messy and are prone to escaping bins without prompting.

Once you've picked out your preferred species, you might ask yourself, how many worms should you buy? Let's take Red Wigglers as an example for simplicity. A good starting point is one pound of Red Wigglers per square foot of bin. That's about 1,000 worms per square foot, as red wrigglers are quite small. If you have too few worms, your food waste will decompose slower - although the worms will reproduce and soon increase their population to match your bin. If you have too many, your worms will slow down their intake and reproduce less, or die off to regulate their population. Either way, your worms will, given time, regulate their numbers to fit their environment. It is preferable to start with too few worms, since you can always purchase more. With too many, you'd need to either wait for the worm life cycle to balance, or to start another bin entirely.

Your worms regulate their own population, and that includes their reproductive behavior. So let's talk a little about the *'worms and the bees'*, to better understand how they do so. Worms are hermaphroditic, which means each individual worm has both male and female genitalia. This is good news for you. Worm reproduction is hands-off on your part; you don't need to worry about female to male ratios. It will always be balanced. From birth, most worm species will take eight to ten weeks to mature. You'll know they're adults when they develop a clitellum, a small ring around the front third of the worm that serves as a vessel to catch sperm during reproduction. It often resembles a band-aid. When two worms mate, they entangle and attach to each other and begin to produce sperm and other reproductive fluids. The clitellum then slips off the receiving worm and forms a protective cocoon in which eggs can develop safely. Each cocoon will result in about five to twenty worms.

Contrary to belief, it is, in fact, possible to mix worm species inside your bin. This is especially helpful if you are expecting temperature fluctuations. Having species of different tolerances means you won't lose all your worms at once if you do forget to move your bin inside before a major overnight freeze. However, if you have several different species, they won't crossbreed, so your population may experience slower growth. Additionally, different worm species cost different amounts of money, and buying different species means buying fewer of each. So diversifying your bin can quickly get expensive. In general, it isn't bad to diversify your worm breeds, but it's not all that much better for your bin, either. It's just a different method.

You should always try to order your worms locally. This guarantees a shorter travel time for the worms and helps promote local, sustainable businesses. Local availability may be a factor in determining which worm variety you purchase. If you're in the **United States**, the best place to get your worms is from Uncle Jim's wormery. You can get 1,000 Red Wigglers for under $50, and they also sell supplementals such as coir and bedding. In the **United Kingdom**, you should look at the Yorkshire Worms company. They have several varieties of worms for every kind of use imaginable, including vermicomposting. **Irish readers** can get their worms at Quick Crop, a gardening site that promotes sustainability and wildlife preservation for home gardens. **Canadian** vermiculturists can get their worms at

Worm Composting Canada, who provide habitat materials for the worms they sell, as well as bins. **Australian** readers can go to Kookaburra Farms, who provide bags of worms and worm eggs, as well as compost bins.

When you buy worms online, there are a few things to watch out for. It's important to ensure that your worms are shipped in a safe, reliable, and humane way. Most farmers will package worms by weight, as it's very laborious to count out 1,000 worms by hand. So you may receive slightly fewer or more worms than promised on the package, but that's completely normal. Make sure the packaging contains some kind of bedding or soil that is pre-moistened. This will keep the worms alive, healthy, and cool during their journey. Next, your worms should be shipped with insulation to protect against temperature changes, especially if you aren't ordering locally and the worms need to be flown for part of their trip. Usually bedding can serve this purpose. If shipping takes more than a week, there should be some food included in the package. This is another incentive to order locally and reduce shipping time. The exterior of the box should be labeled with required warnings such as live animal and temperature advisories. If they are missing, you may need to contact the supplier. Worms are usually shipped in food-safe containers such as Styrofoam and plastic, but sometimes simply arrive in bags. Just like your worm bin, these packages should be ventilated. If the site you are ordering from meets all of these requirements, you can rest assured that your worms will arrive safe and happy.

Chapter 6 - What Should Worms (Not) Eat?

It may seem that your worms are voracious eating machines. They'll eat paper, cardboard, and luckily, even your old food. Is there anything they can't eat? What can and can't you feed your worms? How does their diet affect the conditions of the bin? Can you dump everything leftover on your plate into your bin? Here are the dos and don'ts of feeding your clew of worms.

The primary foods to avoid are animal products. Don't feed them meat or dairy. Worms are herbivores (there's an argument on that – but that's not to be discussed here); they can't process meat or dairy products. Meat in your bin will simply go bad, and that rotten meat will then attract unwanted pests and critters. Dairy is both unhealthy for worms and will probably go bad, well before they can consume it. If you feed your worms dairy, be ready for funky smells after only a few days. This doesn't mean you have to convert to a vegan diet. Just make sure your worms aren't fed any animal products. You also shouldn't feed worms cooked foods. Foods like these are usually seasoned with salt and oils that can really harm your worms. Worms and salt go together as well as snails and salt do (really badly). Oils can stick to their skin and affect their breathing. Oils also make clumps in compost, again making breathing difficult for the worms. Citrus fruits can be given in moderation. They can make your bin too acidic, and in large quantities, can cause acid burns to the worms' skin. Onions and garlic have the same issue: they risk burning the worms if you add too much. Spicy food is very bad for worms; they won't process it well at all. Adding too much spice to your bin could kill your worms. Bread isn't too bad for your worms, but it goes moldy very quickly, especially in the microbe-rich compost bin. If you do add bread, add small quantities and make sure to tear it up. This will increase the surface area, so the worms are more likely to be able to eat it before it grows moldy.

You've seen that there are many food groups that your worms can't eat. Understandably, this can be a little disappointing. You probably started vermicomposting in order to have somewhere to put all your food leftovers, not just the vegetables and fruits. Don't let this deter you from starting a worm compost endeavor. You can still compost traditionally. While the specifics of traditional composting aren't covered here, there are a plethora of resources available. Many vermicomposting communities have more than a few traditional composting members. Your worms' diet shouldn't prevent you from doing your part to make this planet more sustainable.

There are some things that worms will eat that we don't normally consider food. Dryer lint is a nice bedding source. Almost any paper, like tea bags and coffee filters, also work. In small amounts, pet hair can be used as bedding, as long as it doesn't clump. You can also put in foods that take a long time to break down, such as eggshells, banana peels, and coffee grounds.

The best things to feed your worms are fruits and vegetables. Fruit peels, melon rinds, berries, *gourds* like squash and pumpkin are all perfect for your little critters. Fruits high in sugar are especially ideal. You can feed your worms salad leftovers if it doesn't have much dressing. You can also give them frozen vegetables straight from the bag if they're running low on food. You should also supply a steady stream of natural sugars. If you're giving them plenty of fruit, this shouldn't be a concern. If not, try to give them kelp, or something else with natural sugars. Processed sugars are bad for worms, just like they're bad for us.

In the beginning, you can use the quarter weight rule, in which you feed your worms food equal to one-fourth of their weight, to estimate how much food is enough to feed your worms. For example, one pound of worms would be fed a quarter pound of organic food waste. As your bin gets more developed, the worms will reproduce and increase their population. At that point, you can increase the proportion to feed them half their weight. Of course, every bin is different, so it's not a hard and fast rule. Worms don't have teeth, so if you freeze their food and chop it, your worms will be able to process it faster. Smaller pieces increase the surface area of the food so that it decomposes faster. Freezing it expands the water inside, breaking down the cell walls, making the food mushy and easier to eat.

Indoor worms should be fed in smaller, more frequent batches to avoid any chances of bad odors. If your bin is outside, you can give your worms more food, less frequently. For example, for outdoor worms, you can feed them every two or three weeks; and weekly for indoor worms. With a vermicompost bin, less is more. If you underfeed your worms, they'll simply eat the food faster. If they get hungry, they can eat their bedding. If you overfeed them, they'll leave food uneaten, and the bin will attract pests such as flies or maggots.

Chapter 7 - Keeping Your Worms Happy & Dealing With Pests

Once your bin has been running for a while, you'll start wondering whether you're making progress or whether your bin needs maintenance besides the measurable factors. If your worm bin is healthy and thriving, several things should be clear. First, your bedding should look moist. This would mean that your bin is successfully regulating it's moisture content. If it isn't, and your moisture measurements dip far below 70%, don't add water, **add watery foods instead**. You can increase the moisturizing capacity of food by breaking it **down in a food processor** so that it is more of a slurry than solid food. Another sign that your bin is healthy is that worms are congregating around the food you add. If there is too little food, your worms would instead congregate around the bedding and in the bottom of the bin. A happy worm colony will gather around their food supply. You can easily check for this: you should be able to see several worms if you lift the food at the top of the bin. A healthy bin will also have other creatures in it. While these can be deterred somewhat, it's natural and even desirable to have some degree of biodiversity in your vermicompost. As long as you have a lid, those bugs won't escape into your house or garden.

The best indicator that your bin is healthy is mixed bedding. Your bin is balanced if you can see gaps in your compost, the bedding is well spread out, there are plenty of pockets for your worms to move around, and the food is *stratified* throughout.

Indoor and outdoor maintenance require different techniques. Indoor composting is more straightforward than outdoor. Since indoors is practically temperature controlled, it isn't really a factor you need to consider, unless your bin is somewhere semi-indoors, i.e. a garage or covered patio. Your indoor vermicompost needs to be fed gradually to avoid odors, as mentioned earlier. It should also be monitored to make sure it isn't attracting unwanted bugs. One of the best ways to do this is to use a mesh lid, such as cheesecloth. Remember, you'll also want something to catch the draining leachate, or else you'll have gnarly liquid all over your kitchen floor.

If you are composting outdoors, you have more freedom in terms of the size of your bin and feeding schedule. This is especially true if you have a large backyard where you can make vermicompost on a large scale. There are some important considerations before you decide to put your bins outside. First, the bins need to be in the shade. Direct sunlight can dry out the compost and heat it beyond the optimum conditions. Rain is also a concern. An overly moist bin can suffocate your worms. With these two factors in mind, it is almost mandatory to have some kind of cover over your bin. A porch umbrella or an awning is perfect. If you live somewhere that freezes in winter or hits over 90 degrees Fahrenheit in summer, temperature maintenance is your foremost priority. Plan to bring your worms inside for long periods, insulate or cool your bins, and gather extra bedding to help maintain temperatures. If you find your worms simply cannot survive extreme temperatures, you could try seasonal vermicomposting, in which you start your bins approximately ten weeks before these extreme temperatures hit, harvest the castings, and then pause composting until after temperatures go back to a suitable range.

If you notice that your vermicompost is looking clumpy and your bedding isn't well spread, you can mix your bin by hand. Gently agitate the clumps and spread the food and bedding throughout the bin. There is a chance that you could annoy or even injure your worms, so don't do this too often or too aggressively. A hand rake or hand shovel is perfect for stirring your bin. Remember, though, that this isn't regular compost. If it were, regular stirring would be mandatory. In this case, you only need to stir occasionally, if at all.

If, as your bin continues running, production of compost slows down, this can indicate one of several issues. You could be underfeeding your worms, meaning they simply don't have enough material to break down. It could be too cold or too hot, and your worms are getting sick. You could have too little bedding, in which case the castings and mulch are too dense for the worms to move around. You could also have too few worms. The fix for that is simply to wait for them to reproduce or to buy more. Another possibility is that the pH of the bin is too acidic or alkaline. Adjusting feeding and adding lime sources are both great solutions for slow composting.

For a complete maintenance routine and upgrade, follow these steps. Empty your bin and place your worms and compost in a safe place while you work on the bin. There is always the option to purchase a new bin if yours is old or too small. If you aren't going that route, get your drill and refresh all the holes in the top and bottom of the bin. The top ones will probably be mostly clean, but it's likely that the bottom ones will be filled with dirt and will not be draining as well as they could. Re-drilling them will fix this. Next, lay down lots of bedding on the bottom of the bin. Fresh, plentiful bedding will make your worms thrive. You can then add your worms and compost in layers, with bedding in between. This will give your bin a new lease of life. Other upgrades you could make include buying new bins and more worms.

One very solvable problem is worms trying to escape or clumping at the top of the bin. This happens for one of several reasons. The first is that the bin could have become too moist, making it difficult for the worms to breathe unless they are at the top of the bin. Secondly, if their bedding is uncomfortable, such as coated paper or non-organic food, chemicals in these materials could be irritating the worms. There could also be too much citrus or garlic/onion in the food that you're giving them. One key characteristic of worms is that they are averse to light. They like the dark, and when they are exposed to light, they will actively burrow. If your worms are clumping at the top of your bin, and none of the aforementioned issues are present, you can simply place a lamp above the bin, and your worms will burrow back inside the compost.

While biodiversity is healthy for your bin, is it possible to have too many uninvited critters in your compost? While some bugs are a nice thing to have, others may compete against your worms for food or may hurt your worms. Ants can be pesky to have in your house, and a large infestation can be disruptive, but they won't hurt your worms and they help to decompose your compost. One way to deter ants is to use moist bedding and place the feet of the bin in dishes of water. However, if they don't bother you, you don't need to get rid of them.

Beetles can be problematic. If you find beetles in your bin, you may be overfeeding your worms. Underfeed them for a while and the beetles should lose interest. Your worms will be fine, since they will still have bedding to munch on.

Centipedes are some of the *worst* pests to have in your bin. They actively prey on your worms. Centipedes, and their less harmful cousins millipedes, have chitinous exoskeletons, so abrasives like diatomaceous earth can prevent them from invading your bin. If that doesn't work, you'll have to get rid of them by hand, or they will attack your worms. Millipedes are vegetarians and help break down food waste, but they are similar in body structure to centipedes. Millipedes have two legs on each side per segment compared to one on each side for centipedes. If you have one or the other and you have trouble telling them apart, you may want to play it safe and try to get rid of them regardless.

Flies and maggots (the larvae of flies) are generally undesirable. While they probably won't harm your worms, they will out-compete them. Flies have a shorter life cycle than worms, so if you notice maggots, they may threaten to outnumber your worms if you don't take action. Flies are mostly controlled by keeping odors down. Unfortunately, if you notice maggots, then your only option is to refresh your compost pile entirely.

Flatworms, also called planarians, are flat-shaped worms with heads that resemble a crown. These are predators to earthworms and must be removed immediately. Snails and slugs need to be removed. They won't harm your worms, but similarly to flies, they will outbreed and out-compete them. Add a small container, like Tupperware, to your bin and fill it with something yeasty, such as beer, or water mixed with dry activated yeast. The worms won't care for it, but the snails and slugs love it. Once they're in the container, the snails and slugs will drown, and you'll just need to empty out and refresh the container occasionally.

Pillbugs are commonly found in composters and are a rare, desirable critter. They are an indicator of a healthy ecosystem. If you have them, make sure to separate them out from your castings when you harvest, because Pillbugs can be harmful to plants. In summary, if your bin is balanced, correctly fed, and monitored frequently, you shouldn't need to be too concerned about harmful pests.

Chapter 8 – Harvesting & Using Your Vermicast

Vermicast (Vermicompost) is made of the excrements of worms as they eat and digest organic food waste. Their excrement is a potent fertilizer, and also contains bacteria, plant matter, and humus. Humus is the dark, nutrient-rich component of soil, and it's the primary product of compost. It contains essential nutrients such as nitrogen and carbon in high concentrations.

When your bin is ready for harvest, there will be several signs. First, you'll notice that your worms have slowed down. They'll reproduce less and eat less. This is an indication that there is too much vermicast in the bin, and the worms are uncomfortable. Red Wigglers will seem especially slow since they and several other compost breeds don't burrow into soil. Second, your worms will be smaller than usual. When worms can't reproduce and grow, they physically shrink in order to ensure that each worm has something to eat. Third, the compost itself will have several visual changes: it will look dark brown, like wet soil, even when the moisture content is correct; it will look soft and smooth, like velvet; and finally, the texture will be uniform, without many visible chunks of food (although, if you chop up the food before feeding your worms, it may always look uniform). If your worm bin is showing these signs, it's probably time to harvest your castings! *As a rough guide, harvesting will generally be necessary every 3-6 months, depending on how much you feed them, and many worms are living in your bin.*

Harvesting castings can be messy. You need to separate the castings and humus from the worms. Unless that is, you don't mind burying some of your worms in your garden and just order replacements. Separating the compost is actually easier than that. It may seem undesirable to sift through worm leavings and pick out a thousand worms, but there are several methods to make this job easier. First, you can just use your hands. It's slightly tedious and time-consuming, but you can enlist kids and family members to help. For this method, you can use the worms' distaste of light to your advantage. Pull manageable piles of your compost out. With a flashlight, scare the worms into the center of the pile. Then, scrape the castings off the sides of the pile and collect them until you hit more worms. Use the flashlight again. Rinse and repeat until you have two piles, one of the worms and one of the castings.

You can also use worms' love of food to draw them away from their castings. After the worms have eaten their food, stop feeding them for a few days. They should finish up all the smaller bits of food in the bin. When it looks like they've moved onto their bedding, put some of their favorite food at the top of the bin in a corner. Something like sugar-rich fruit is perfect. The worms will all congregate around the new food, and you can easily separate out the vermicast.

The least labor-intensive way to separate your worms is to use a sifter. As we discussed in Chapter 3, options for these vary from thousand dollar industrial tumblers to kitchen strainers from the grocery store. You can even make your own sifter with some wood and window screen. Make a square screen bordered by wood, with a bigger box underneath. Put your compost on the top screen and shake it. The compost should drop in to the box underneath, and the worms will be left on top. With the strainer method, simply scoop out a small sample of compost, shake it in the strainer over a bucket, and you'll be left with some worms and a bucket of rich compost.

It's finally time. You've cultivated your worms. You've been feeding, watching, measuring, and managing for weeks. You just sifted through a thousand worms to extract their nutritious fertilizer. Now, finally, it's time to use your compost. Wait. How?

There are dozens of ways to use your compost, worms, and worm tea. At the most basic level, you can use the vermicast to fertilize your plants. Take the vermicast and spread it so that it covers the soil at the base and underneath the leaves of your plants. Pat it down and build it up until it's about one-half to one inch thick. This works wonders for both indoor and outdoor plants. You can also add it to soil before planting. When you prepare your garden for planting, turn your soil. A more experienced gardener is likely already familiar with this process. Cover your unseeded garden with two or three inches of compost. Then, take a rake or shovel and mix the compost in with the soil. When you are ready to plant, your greens will have incredibly nutrient-rich soil to grow in.

You can also use your compost as a seed starter. Some seeds need a little extra care when they first start out. You can plant them in an old ice-tray, using your compost as soil. Once they start growing, you can gently transfer them to your garden. If you've mixed your compost into that garden as well, your plants will certainly be grateful. If you have a large amount of vermicast, you can spread it thinly onto your lawn during the summer to encourage your grass to grow greener, faster. On the other hand, if you don't have plentiful amounts to spare, you can use it on patchy parts of your lawn to help make your grass more uniform. For this application, a seed spreader from a hardware store is perfect for helping you get a thin, even layer with easy effort. You can also use your compost as a mix-in if you have dry, dead soil. Maybe you or a friend has an overused plot of land that could use a break. Mix it in like the garden and then leave it, repeating every other week, and watch as the dead soil comes to life. But don't give your vermicast away too frivolously – as remember, it's worth up to $400 per cubic yard!

There are also several uses for worm tea. To make worm tea, steep your castings in water for 24-48 hours, as described in Chapter 1. Remember that leachate is not worm tea, and can't be used in the same way. The most popular use of worm tea is to spray it directly onto the leaves of plants. This lets the plants absorb the beneficial nutrients quickly. Worm tea is an excellent external pesticide, so if you have plants with holey leaves, give them a spritz of worm tea. You can also water your plants with it. You don't need very much for each plant in order to start seeing benefits. In an incredibly unique and sustainable application, worm tea can be used in hydroponics to enhance the growth of plants. Hydroponics is a method of growing plants without any soil, in which the plants are suspended in water containing dissolved nutrients. Using worm tea to enhance the liquid solution in which the plants grow is an excellent way to facilitate their growth.

Any garden would be lucky to be blessed with worm compost. Wherever you use it, your plants, trees, grass, or flowers will thank you. Your vegetation will grow faster and healthier. They'll be less susceptible to bugs. They'll be greener, too. You can even find more uses for the worms. If you have some extra worms between harvesting cycles, you can use them as fishing bait, or reptile food. Red Wigglers are especially good for bait, and they get their name from how they look when hooked to a fishing line. You may even find you still have additional compost after all this. In later chapters, you'll learn how to convert your extra compost into your local currency!

Chapter 9 - FAQ's

Before we go on to discuss common beginner mistakes, the business of worms, and how you can make worm composting fun for all the family; let's quickly address some frequently asked questions. The topics range from other materials your worms can eat, what to do if worms are dying, and how to deconstruct your worm bin if you ever want to move on from this hobby. When you do begin worm composting, you should keep this chapter bookmarked or saved so that you can come back to it whenever you find yourself asking questions.

How much will this cost?

It's completely up to you how expensive vermicomposting will be. You can hop online and grab a $300 compost bin if you want to. Alternatively, you might decide to go for the cheapest bin possible. You'd buy supplies from the hardware store, get your worms locally, and use items from your own home whenever possible. A thousand red wigglers will cost about $50. A storage bin from a grocery or hardware store will cost about $15, and a five-gallon bucket usually costs less than $5. If you're using an old garbage can, that would be free. Also free is the food waste and the bedding. You can ask friends for supplies, extra newspapers, and coffee grounds. You can also get neighbors involved if you aren't producing enough food on your own. A knife is less than $10, and if you want to splurge on a drill, that will set you back about $40. Overall, you could invest less than $100 to start your vermicomposting journey. Again, it's very easy to surpass this number, but for most, worm composting is an incredibly affordable and sustainable hobby.

Can I feed my worms waste from my pets?

In general, you should avoid feeding your worms your pets' waste. Since worms are vegetarians, feeding them waste from carnivorous pets is harmful to them. Cat and dog waste should not be fed to worms. This is not true for all pets. Herbivorous animals do produce waste that worms can eat. Horse leavings specifically are a favorite of compost worms. The waste of sheep, cows, rabbits, and vegetarian rodents are also okay for your worms. If you have a rodent at home, a guinea pig, or hamster for example, you know that cleaning their cage can result in lots of wood chips, waste, and leftover food going straight into the trash. If that animal isn't eating animal products, you can put all of that waste into your vermicompost bin. This will serve both as bedding and as a food source for your worms.

Can I feed my worms compostable diapers?

If you've invested in some compostable diapers, you may be wondering why you can't just put them in your worm bin. Worms will be able to break these diapers down. However, it's important to note that these diapers will act as food, not bedding. Your worms will eat it first. Because of this, you need to be careful not to overfeed your worms. Only add a few diapers per week. You can experiment with this to find the right number for your bin. In general, you shouldn't be using your vermicompost as a regular diaper genie. Additionally, worms (mostly herbivorous) can't process meat products. Similar to the pet waste rule, if your baby, or more likely your toddler, is eating meat, you shouldn't give their diapers to your worms. You should also be cautious if your baby is exclusively drinking milk or breastmilk. Monitor your worms, and if they grow sluggish, avoid the diaper, or if it starts to change smell, that's a sign the worms can't eat the stinky stuff in the diaper.

Do I need to hire a worm-sitter when I go on vacation?

If you're planning to leave home for less than a month, you don't need to worry about your worm bin. Worms will eat their bedding when their food runs out, and bedding lasts for a long time. Before you leave, make sure to stock up on bedding, especially the moisture-absorbing, long-lasting kind. Coir and cardboard are great for their long-term lifespans and water-absorbing capacities. Add plenty of bedding before you leave. Move your bin to a safe area or even indoors so that it won't get too hot or cold. Make sure it doesn't get too moist by guaranteeing that leachate will have a place to drain. Put a cloth over the bin to prevent unwanted insects from invading while you're gone. If you'll be gone for longer than three or four weeks, ask a friend to check on your worms every couple of weeks.

Is Worm Composting seasonal or year-round?

Your worm bin will produce vermicast year-round. You may notice worms slowing down in colder months as their metabolism slows to save energy. However, this isn't complete hibernation, and they'll still be eating what you feed them. You may need to feed them less in winter as they slow down, but you'll still have compost year-round. Since your garden probably isn't as active in the winter, you probably won't need many castings. You can feed your worms mostly bedding during winter.

The worms will keep working, and by spring, you'll have a healthy, hungry pile.

What if all my worms die?

Walking over to your bin only to discover no sign of life is a devastating moment. Unfortunately, it can happen. If your bin is controlled for physical factors and your worms seem healthy, you shouldn't have a die-off. If you do, it's very likely because one of the factors from Chapter 4 has occurred. Maybe your bin got too acidic, maybe it froze, maybe it got too moist, and the worms suffocated. Maybe the bin was invaded by a predatory pest. Either way, while this is definitely disappointing, don't be disheartened. The most important thing about making mistakes is to learn from them. Troubleshoot as much as you can, consult with an experienced vermiculturist, order some more worms, and try again. You are certainly not the first worm bin owner who has ever had to do this – and you won't be the last!

What do I do with dead worms?

Dead worms aren't automatically a sign of something wrong in your bin. Worms live for about a year, and up to four. When they die, their bodies decompose and are added into the rest of the compost. If you notice a few dead worms here and there, you don't need to take any action. This is simply an indicator of the life cycle of your worms. This is true even if your bin is relatively new. You may have simply purchased older worms at the time of purchase. If you notice large numbers of worms dying off at once, this could indicate something more sinister. Troubleshoot using what you've learned in the book so far, and if you still have worms dying, transfer the living ones to a new bin and start over. It's much better to save your healthy worms and lose your compost than to save your decomposing food waste but kill all of your worms.

Why are all the worms at the top of the bin?

If you notice worms congregating at the top of the bin, or even on the sides, this is usually a sign that something is off balance. First, check the weather forecast. Worms can sense when it's going to rain through changes in air pressure. When they sense rain, they move upwards. This instinct prevents them from drowning when the soil gets soaked. They'll go back down once the weather returns to normal. If it rains for a long period, don't worry. They won't stop eating while they're at the top. If it's not going to rain, then there is likely a problem with the bin. The worms could be having trouble breathing, there could be a pest infestation, or there could be too much citrus and other bad foods in the bin, making the worms sick. Once you fix the balance of your bin, the worms should make their way back inside. If not, or if you need to rush them, you can use a flashlight to chase those worms back where they belong.

All my worms are on the bottom of the bin; what should I do?

Your worms will primarily move towards food. If your worms are congregating on the bottom of the bin, that just means most of their food is also at the bottom. This also means that the worms are able to get to the bottom easily, so at the very least, it's an indicator that your moisture levels and bedding amounts are good. Stir the pile gently with a shovel or hand rake, and try to create layers inside your bin. If you alternate bedding and food, then you should see your worms travel up the bin as they eat. They'll start at the bottom, and work their way up, eventually favoring the top of the bin, since that's where you're adding new food. Another perk of this method is that when it's time to harvest your bin, most of the worms will be at the top, so you can peel back the top layer and simply scoop out that nutrient-filled soil.

What if my bin gets moldy?

Normally, your worms will be moving around and eating enough that they don't give mold a chance to develop. If you have mold in your bin while there aren't worms in it, you shouldn't be surprised or even concerned. Stirring it around will take care of it. If you have worms in your bin and mold still makes a way in, this is a strong indicator of one or two issues. You could be overfeeding your worms. If the worms can't eat food faster than it breaks down, that food will get moldy and nasty. Another indicator of overfeeding is unpleasant smells, so if your moldy bin is stinky, it's very likely because of overfeeding. Alternatively, your bin could be too acidic. Mold favors acidic environments and foods. As long as the pH is still between 6.5 and 7.5, this isn't a huge issue. There's a strong possibility that even when your bin is acidic enough to attract mold, it won't be acidic enough to harm your worms. Calcium additives like diatomaceous earth, lime mix, or even eggshells all help bring the pH of your bin back up to normal if it's below 6.5.

If I cut a worm in half, do I now have two worms?

It sounds like a fun science experiment. You cut a worm in half, and both are now wiggling around like normal. Except, this old myth doesn't translate to reality. On occasion, you can get lucky with your cut, and avoid chopping anything vital off a worm. In that case, you'll have a shorter worm and its dead tail. If you don't get lucky, all you'll manage to do is kill the worm. There are plenty of fun science experiments to do with vermicompost and with kids, but cutting worms in half isn't one of them. If you handle a recently deceased worm and it seems to be wiggling on its own, this is because salts on your hands are reactivating the electrical signals in the worm's nervous system and causing the muscles to spasm. It's not wiggling because it's alive. This is actually a great science experiment if your kids are learning about the brain, nerves, or electricity at school. You may even remember doing this same experiment with frog legs in a biology lab.

Should I add my worms to the bin immediately or wait a week for food to decompose?

You may have heard advice that you should start your bin before you get your worms. This is a bit of a misconception. When you first add your worms to your brand-new bin, they will seem very sluggish and may not eat much for the first few days. This is not because the compost isn't ripe enough. The worms are living organisms, and they experience confusion, as many other organisms do. This is also why you shouldn't stir the bin very often. It can easily disorientate the worms. When they go from a farm to a small box or bag that moves a whole bunch, to a big bin with lots of yummy stuff, they're likely to be pretty disoriented. It's normal for them to take a few days to adjust to their new home. If you didn't know this, it could seem like they were waiting for the food to break down a little before eating it. They aren't. They'll start eating when they're ready. It's better to start your compost and worms at the same time, and avoid all those smells and mold that come from a stagnant bin.

Can I add worms from my garden to my bins?

In general, garden worms are not suited for compost. Most worms found in gardens are *endogeic*, meaning they burrow near topsoil, or *anecic*, meaning they burrow deep into the soil. Composting worms are *epigeic*. They don't burrow at all. This is why bedding and air space are so important; your worms won't be able to move around in their bin if it's too dense. The worms you find in your garden won't be comfortable in your bin. It may seem like a cheap alternative to purchasing worms in bulk, but you can't move worms away from where they're happy. You also shouldn't add too many worms from your compost to your garden for this same reason.

What if my pets get into my compost?

Even though your bin has a lid and is probably on a stable surface, there's still a chance that a curious companion could knock the lid off or tip the bin over. If your dog or cat has snuck a few bites of your vermicompost, you don't have too much to worry about. Most of the food you feed your worms is okay for your pets to eat too. There may be a few food products that worms eat, but pets can't. For example, if your dog eats a handful of grapes, it's definitely time to call your vet. Additionally, if your pets get into the bedding, you might want to check to make sure they're okay. Cats and dogs aren't very good at digesting cardboard. If they ate a few worms, however, they'll be fine. Overall, if you're feeding your pets the correct amount of food, they shouldn't be very interested in a bucket full of soil and worms. If you are doubtful, call your vet. Better safe than sorry.

How do I introduce the worms to my bin?

Start slowly. Empty your bag or box of worms into your bin and let them spread out on their own. Remember that sluggishness is normal. Try not to turn or stir your compost just yet, as this will only further confuse your new worms. You also don't want to feed them very much for the first few days. They eat less in their stressed state. They won't get too hungry, don't worry. They'll eat their bedding and the little food you give them in small quantities until they can get their bearings. You should also put a lamp near the bin. Your worms will be exploring their new home, and a lamp will keep them from exploring outside of it. After about a week, you'll notice your worms moving faster and eating more. At that point, you don't need to be as cautious with them.

What if I receive my worms before my bin is ready?

If their new home won't be ready for a few days after your worms arrive, you can keep them in their packaging. It should be designed to last longer than the predicted shipping time. If you need to hold off longer than a week, you can store your worms in a new box similar to the one in which they arrived. As long as it has the correct moisture content, lots of bedding, and a bit of extra food, your worms will be okay. You may see advice that you can refrigerate them until your bin is ready. This is incorrect. While it is okay to refrigerate bait worms, your compost worms cannot tolerate temperatures under 50 degrees Fahrenheit. A Tupperware container or a baking dish with a cover are perfect storage devices for your new wigglers. Just make sure the cover has ventilation for the worms to breathe.

What if I forget to feed my worms?

Your worms don't need to eat as much as it seems. While they'll probably get through the pile of scraps you give them once a week, this is simply because they prefer fruit and veggies over bedding. You would too, if your options were apple or paper. As long as your bin has plenty of bedding, your worms will be okay with a missed feeding or two. If they don't have enough bedding, however, and they run out of food, they will start eating their castings. This will make them sick, and they will likely start dying off if you don't feed them. They may also start trying to escape in an effort to find better food sources. And, they'll probably stop breeding as they focus on survival. So, while you don't need to panic if you forget to feed them for a week, you shouldn't make it a habit either.

What do I do with leachate?

Since leachate is not a byproduct of worm digestion, it does not have the beneficial microbial properties that worm tea and castings do. Leachate isn't very useful; not nearly as much as the other products of vermicompost. It can be used to water plants, but you need to do some prep work. First, if the leachate smells weird, don't use it at all. Throw it away. It's rotten and will only hurt your plants. If your leachate smells fine, then you want to address the main issue that makes it undesirable. Having seeped all the way through the contents of your bin, leachate doesn't have much oxygen in it. This is why it's so important to have it draining from your bin, as it can suffocate your worms. Unless you oxygenate it, it will do the same to your plants. You can leave it out, uncovered, for a day or two, or you could use a fish tank bubbler, or you could replace your arm workout for the day and shake it in a sealed container until it's foamy. Once you've oxygenated it, you can spray it on your plants, just not the ones you plan on eating.

What if my leachate isn't draining?

Make sure your leachate has room to drain. Prop your bin up with a brick, wooden blocks, or another small but sturdy object. Make sure there are holes in the bottom of the worm bin, and that the holes aren't clogged. If you have a pre-made bin, make sure the spigot is always open. If your bin still isn't draining, make sure the compost on the bottom of your bin isn't plugging the holes. That compost on the bottom could also be too dense for the liquid to seep through. If you've addressed all these problems and you still aren't seeing leachate draining, check the moisture levels of your bin and the conditions of your worms. If everything else looks okay, your bin simply isn't producing that much leachate. The compost and bedding are absorbing it. This is actually great news because it means the balance of your bin is just right.

Do worms have eyes?

Even though worms are sensitive to light, and you've probably used that to your advantage by now, worms do not have eyes. Instead of eyes, worms are covered in several sensor cells all across their bodies. These receptors transmit information to the worm's brain, which lets them know that light is present. Worms need to be sensitive to light, because being out in the sun for too long can dry their skin. Since worms breathe through their skin, drying out means they suffocate. Excessive exposure to light can lead to paralysis and eventually, death.

How do worms eat if they don't have teeth?

Worms crawl and burrow primarily by expanding and contracting their muscles. They eat by clenching the muscles near their mouths. Microbes help them to soften and break down their food so that when they finally go to eat it, it can be easily swallowed. The food moves through their body, again driven by muscles pushing it through. Worms don't have a stomach, but they do have intestines. As food works its way through their digestive system, some of the nutrients are absorbed by the worm. The rest are passed out the other end in the form of castings.

How long will my worm bin last?

If you are taking proper care of your bin, it will last for as long as you are feeding it. The worms will reproduce on their own, so you shouldn't need to buy any more, other than to replace any that die naturally. It is good to get into the habit of regularly cleaning your bin. Doing so will keep your compost more healthy and fresher. You should also make sure to regularly clean out the drain in your bin. Whether it's a spigot or holes which you drilled, it will need occasional cleaning. The best time to clean your bin is when you are harvesting the vermicast. You are already pulling everything out anyway; you might as well also clean the bin. You could also make two bins for composting, but only have one in use at a time. That way, when you are ready to clean one, you simply have to transfer the contents from one bin to another. This is a nice way to save time when harvesting as well. As long as you are cleaning your bins, regularly harvesting, and replacing the bedding, your worm bin will just keep on going.

What will be the biggest challenge I face when I start vermicomposting?

This question can be really subjective.

Some people find handling worms to be the hardest part. Some find harvesting castings too meticulous, while others find it theraputic. I have found that for most people, the hardest part is waiting. A lot of vermicomposting is waiting. You wait for the worms to arrive, for them to get settled, for them to start eating, for the castings to start piling up. You wait for the compost to be ready, and then you have to wait through several stages of harvesting in order to get your fertilizer. You wait for your plants to grow or your grass to come back in color. You wait for your worm tea to brew, and then you wait for your plants to get stronger. When you first start composting, you're going to be so excited to see the results. But you have to wait. If you persevere, you will reap the rewards.

What if I decide I'm done vermicomposting?

We'll be sad to see you go. Vermicomposting is a rewarding experience for you and for the planet. But if you can't continue, there are a few steps you can take to ensure your materials don't go to waste. First, you can sell or give away your worms. You can give them to a community garden or a local bait shop. You could give them to a fellow worm composter. You could encourage a friend to adopt your bin. You can also give your worms to a company that sells them. You can clean your bin and repurpose it, provided the holes are sealed. As long as you aren't storing liquids in it, that bin can be used like any normal storage bin. You can take the leftover compost and bedding and spread it over your yard or garden. Many bedding materials can be used in gardening. Paper is a sustainable alternative to plastic weed nets that are usually used for landscaping. Yard clippings, fur, and lint can all be put out as well. They'll probably be taken by a nesting bird or squirrel. Some items, like the lime mix, may not be easy to get rid of, but ask around your neighborhood or at local hardware stores. There will be someone else who could use it. The best option is adoption. Give your setup to a friend who is interested in vermicomposting. That way, the practice lives on, and your friend has an easier start.

Chapter 10 - The Business of Worms

Vermicomposting is a satisfying hobby. It feeds your garden and your soul. And if you have some leftover worms or compost, it can feed your wallet as well. Making money from your worm composting is definitely easier said than done, but with dedication, it can be quite lucrative.

If your worm bins are constantly breeding, you should look into selling those extras. You can expect to receive about $40-50 per pound of worms. The market for worms is much more than just composters, too. Farmers, chicken and reptile owners, fishers, and bait sellers are all looking for worms to buy. Worm tea and your homemade fertilizer are worth a good amount of money. For example, worm tea on amazon can go for upwards of $40 per gallon. Castings are typically worth $400 per cubic yard.

But how do you start your own compost business? It's similar to starting any other business. You need to invest in equipment and supplies, produce on a large scale, market your product effectively, know how to sell it, and ship it efficiently. If you are dedicated to going large-scale, you're also going to need more space. There are several steps involved in starting a business. The first is mental. Running your own business is a lot of work. It will most likely be a big time and financial commitment, so you need to have savings stowed away, and be prepared to spend them. If you have strong time-management skills, leadership skills, lots of perseverance, and are self-confident, you have the skills needed to begin an entrepreneurial endeavor. I also recommend reading some self-development and business books to get a better idea about different approaches to sales, marketing, and accounting. These principles can be applied to any business. Also, speak to others who've built successful agricultural businesses and attend any local conferences, seminars, or networking events to meet other successful business owners in the industry. At the end of the chapter, I'll give you an example of a conference you could attend, if you're based in the USA.

Once you get over this mental hurdle, there are some legal obligations to be aware of. When you start a business in the US or the UK, as well as in many other countries, you will mainly have two options: You can form a corporation, or you can be a sole proprietor. If you are a sole proprietor, you will have a simpler time establishing your business. You won't need as much paperwork, just general licenses. You won't be subject to extra taxes. You, personally, are in charge of the finances of the company. This aspect can be good or bad. If the business is in the black, you probably don't need to worry about this too much, and you could even see benefits on your tax returns. If your business is in debt, however, you are personally responsible for that debt. If your business needs a loan, that goes in your name, and your assets are vulnerable if you default upon that loan. It's also harder to establish credibility or accrue investors. Since a sole proprietorship is riskier, people are typically less trusting of them.

On the other hand, a company, such as a limited liability company (LLC) in the US, or a LTD company in the UK, requires more work upfront but is more stable and trustworthy. There are different types of corporations to choose from, but broadly, using a corporation means your business will be disconnected from your name. Just because your LLC might be in debt, doesn't mean that you are personally liable for that debt. This method takes lots of paperwork and some extra taxes, but the validation from the government registry means your business will look much safer to any future investors. Whichever type you choose is completely up to your own situation and preferences. Make sure you put thought into this decision; it is very important and will impact most further aspects of your new business. On a small scale, no licenses are needed to start worm composting.

Once you have your company set up, you can move on to production. There are a few different methods for mass-producing vermicompost. You could go the low-tech route, which is cheaper but more labor intensive. You could also go the high-tech route, which is more expensive, but easier. The low-tech method involves windrows. Windrows are piles of vermicompost arranged in long rows. These rows are usually 10 feet long, 3 feet wide and 3 feet tall, but can be smaller. One of these rows is approximately 5 cubic yards of compost, which can make anywhere from $150 to $450 each, come time to harvest. You'll need a pound of worms per square foot, and half that much weight in food waste. They need to be turned and formed with large appliances like manure spreaders or compost turners. Feeding the windrows will also require industrial equipment. While this seems pricey, it is still the cheaper of the two options. It is also very labor intensive, so unless you are willing to make production your full-time job, and then some, consider the second option.

Hight-tech (option 2) involves using a piece of industrial equipment: a Continuous Flow Vermicomposting System. These machines turn the compost for you. The bottom is made of mesh. A mechanical arm runs across the bottom of the pile, sifting the castings into a container without disturbing the upper layers of worms. They are low-labor but expensive, upwards of $5,000 if you want the Michigan SoilWorks CFT. That model outputs 160 pounds of compost for every 200 pounds of waste put in. There are several other models of machines. Michigan SoilWorks is one of the leading brands for CFTs, and they have several options for large scale production. These machines also lend themselves to automating other aspects of the process, such as feeding, adding bedding, and even packaging.

One thing to be aware of is that when you produce compost on a large scale, you need to compost the food before you can give it to your worms. This is done through traditional methods of composting, including heat cycles. This is required because of a regulation called the

'Process to Further Reduce Pathogens'. The Environmental Protection Agency in the United States mandates that pathogens be brought down to acceptable levels in order to prevent the buildup of viruses, salmonella, and fecal coliforms, amongst others. Once the compost has reached a specified temperature, it can be declared safe and you may begin to use it. There are also laws that vary region to region regarding how to mail live animals and animal products. Contact your local authorities before starting your business, as violating these laws is an easily avoidable but very serious mistake.

Now that you have a product to sell, you need to pick a target market. You can't sell to everyone; you'll spread yourself too thin. There are three basic markets for a worm composting business. You can sell and produce directly to gardeners, to farmers, and to others who want to buy worms or compost. A good example of someone who wants to buy worms, but isn't interested in compost, is a fisherman. Fishing is a huge market for worms, specifically Red

Wigglers. Don't restrict your business to just composters.

You could also indirectly produce, but directly sell, also known as reselling. By purchasing worms or compost, then adding value to it, you can make a profit without investing in large-scale equipment. You can also add value by selling starter kits, or including instructions, lesson plans, or other materials. You could add something unique, like worms packaged with your own compost as a starter material. There are tons of options. The important thing is to realize that this method requires lots of creativity. If you're creative and on a budget or don't have the money for a large farm, the indirect method can still make you some extra cash. Finally, you could run things from a marketing perspective. Use your creativity to help other local farmers advertise their products. You could also establish a brand and then have other farmers sell under your brand, for a shared profit. You could also market in the most direct way possible: sell your extra worms, compost, and worm tea to other avid vermiculturists.

The line between picking a market and restricting your options is thin. Look to other suppliers, like the ones mentioned in Chapter 5, as examples. What is their target market? Whom are they primarily marketing to? Which business category is their main focus? You could even contact some of these sellers directly for advice. There are even organizations and conferences you could attend, such as the Worm Farming Alliance and the North Carolina State University Vermiculture Conference. Social media groups and pages are a great source of community where you can find more advice and support.

Chapter 11 - Worm Composting for Children and Educators

Worm composting is a wonderful hobby for you and your kids. Children can learn about taking care of animals, recycling, gardening, and they'll get some hands-on experience as well. If you have a particularly squeamish child, this is also a great way to introduce them to gentle creatures that can't bite them. Since vermicomposting is easily scaled up, this is a great project for educators as well. A classroom worm compost bin will lend itself to harvesting two or three times over the course of a year. Additionally, since the worms are okay if left unattended for a few weeks, a bin left alone for school breaks will be just fine when pupils get back.

Making your farm child-friendly doesn't require changing much. Make sure your farm is on a stable surface that's low to the ground. It should be low enough that your child doesn't have to tiptoe or stretch to see inside. This, along with a stable surface, will keep your bin from tipping over. You can also make a window in the bin. Simply cut a piece of the side off and replace it with a few layers of plastic wrap, secured with tape. This will let kids see the layers of the bin, and even how the worms like to move around. To encourage worms to come close to the window, tint the plastic wrap red. You can do this with markers or buy it pre-tinted. The red filter will mitigate the intensity of light, and the worms won't be repelled from it like they normally would. Make sure to keep your supplemental tools out of reach. Shovels and other hand tools can be sharp, and handfuls of calcium aren't the safest thing for your kids to be getting hold of. The last step you'll need is to secure the lid so that your kids can't open it on their own. If you have a plastic lid, punch holes on either side and tie the lid down, with the string wrapping around the bottom of the bin. If you have a cloth lid, use elastics to secure the cloth, then cover those elastics with heavy-duty tape. Sealing the lid makes it so your kids can't dig around in the bin or accidentally leave the lid off for an extended period of time.

You can also make your bin more child-friendly by getting them to decorate it before you fill it. Stickers, paint, anything they like. It's an easy way to incorporate arts and crafts, giving them a sense of ownership over the project before it even starts.

There are several activities you can do to help children learn while they take part in worm composting. The first step is to teach them about core concepts such as recycling, sustainability, ecosystems, and basic gardening. YouTube videos, children's shows, and books can all help them learn these things. Then, introduce them to the worms. Use a hands-on approach. Unbox the worms together and let them hold them. Encourage the kids to be gentle. Explain the journey the worms have been on: the travel from farm to truck, to your home or classroom. Teach them about what worms eat and even how they do it. Teach them about layering the bin, what bedding is, and how the worms move through the bin. Let them watch for a while as the worms settle into their new home.

The most important factor to keep in mind when teaching young children about science is that active learning is best. Let them use their senses to make observations. Encourage questions. Ask them questions, as well. If they are old enough, let them use their hands to explore the worms and the compost. Have them feed the worms themselves by placing fruits or veggies into the bin.

You may have kids or students that are less than excited about playing with worms. Don't be discouraged. Teaching kids to be comfortable around animals is a valuable lesson that they'll carry throughout their lives. For hesitant kids, you can start by introducing them to the other aspects of vermicompost. Have them be your helpers in shredding paper, gathering food, making the box, and even picking out the place you put the bin. This will give them a feeling of participation and ownership. When they feel involved, their curiosity will increase. Especially if you have other children or classmates involved, that curiosity will help them grow more comfortable in handling the worms. When they feel a little more ready to handle the worms, you can help them by holding the worm first, letting it crawl onto their hand, or holding your hand on top of theirs. That extra sense of safety will help kids adjust to the idea of holding something that scares them. Once the worms are in their hands, the kids should learn how safe and fun it is to hold them, and their fear should go away. The key factor here is never to force the kids to do something they are afraid of. These steps are designed to help them overcome fear. If you simply pick up a worm and place it onto their hand before they feel ready, they may simply become more afraid, since now they are both uncomfortable and have had control of the situation taken away. If, after all this, your student or child is still unwilling to interact with the worms directly, don't worry. There are plenty of lessons to learn and ways to participate without it.

Vermicompost isn't just a teaching tool for small children. Pre-teens and teenagers can get a lot out of it, too. Especially right now, it's incredibly important to teach the younger generation the principles of sustainability. Teaching teens how to respect the environment is imperative. Worm bins can teach them about ecosystem interactions, food webs, biomes, and all kinds of earth science and gardening principles. Gardening has also been proved, several times, to boost mental health. If you are a parent of a particularly frustrated and angsty teen, gardening may help them to relax.

One of the most satisfying feelings I know is teaching something to someone and watching them learn and grow. Writing this book is itself a form of that. When you teach kids and teens, and even adults how to compost, I hope you feel that same swell of pride. It's very important to anyone's learning to have hands-on experience.

Environmental education is rooted in exploring nature and all its wonders. When you teach a classroom of kids how to compost, you are giving them this gift.

Chapter 12 - Tips for Success & The A-Z of Worms

What are some other ways you can not only troubleshoot your worm bin but improve it altogether? Maybe the business side of things isn't for you, but you still want to produce more compost. In this chapter, you'll learn how to bring your vermicompost to the next level. You can also use this chapter as a reference for any vocabulary that might have left you scratching your head throughout the book.

Perhaps the best way to grow your vermicompost is to take a few of your extra worms and add them to a second bin. Splitting your bin has a variety of uses besides population control. If you aren't ready to harvest yet, but your bin is starting to fill up with compost, you can split the bin into two, and add new materials to each. Not only will your worms increase their population to match their new home, but the castings will also be shared between the old and new bins. This means that you can wait a little while longer before harvesting. Each bin will be less crowded. Less crowding also means your worms will be more willing to eat and reproduce. They'll have less competition with each other over who gets what food. Splitting your bin can cause a population boom for your worms, and in a few months, they'll be ready to be split again. If you find that you have tons of food waste, and your worms can't eat it fast enough, adding a second bin quickly can solve that problem. Two bins mean twice as much compost, once the population's balance. Splitting your bin can also be a nice way of helping friends and family start their own. It's much easier to get into vermicomposting if you're given a bucket that's ready to start eating.

Like learning anything, you're going to make mistakes. Some common ones are fairly easy to address. First, make sure you're feeding your worms the right food. It's easy to forget what they can and can't eat when you first start out. Make a table of good and bad foods and post it somewhere noticeable like on the bin or on your fridge. This will help you and your family avoid any mishaps. Another easy mistake is choosing the wrong species of worm. This is why scientific names are provided in Chapter 5. If you use the technical Latin names for your worms, you shouldn't run into any issues. Buying worms from a compost-specific site also helps to ensure you're getting the right kind. A common issue that people forget to address is cleaning the bin. Keeping the bin clean between harvests will keep the bin free of harmful pathogens and microbes. Additionally, regular cleanings can help prevent clogging the drain holes or spigot at the bottom of the bin. There's a chance your worms aren't ridding every single nook and cranny of the bin free from food scraps. Cleaning also prevents hidden pieces from smelling or attracting undesirable pests. Remember always to keep track of those three conditions in your bin: acidity, temperature, and moisture. Forgetting any one of these is a recipe for disaster. Finally, remember to do pest management whenever you can. Observe your bin for any critters that are in there uninvited. They could be stealing food from your worms or even actively attacking them. Periodic checks can help ensure your worms are safe.

What follows is a vocabulary of key terms that can help you further expand your vermicomposting expertise. Also included are descriptions of several different species of significant worms and insects. This is another section that will be useful to bookmark for future reference.

Acidity: The measure of how acidic or basic (alkaline) a substance is. Acidity is measured on the pH scale. As the pH measurement of a solution goes down, it is increasingly acidic; as the measurement goes up, it is increasingly basic. Your vermicompost should be between 6.5–7.5 on the pH scale.

African Nightcrawlers: (*Eudrilus eugeniae*) Large worms, good for composting and fishing. These worms can easily grow over six inches long. They are accustomed to temperate climates of 60–90 degrees Fahrenheit. There are approximately 200 worms per pound.

Anecic Worms: A group of worm species that have deep burrows. Not ideal for composting, these worms are great additions to garden soil for their ability to aerate and add nutrients to deeper plant roots.

Bait Worms: In fishing, worms are used to entice fish to bite onto a hook. Commonly sold in bulk, these worms are sometimes the same species used to compost.

Bedding: A filler component of composting bins. Bedding is made up of compostable but long-lasting organic materials. It insulates bins, decreases the compost's density, and gives your worms a reliable source of food.

Bins: The container in which composting takes place. It can be handmade or store-bought. These always contain air holes and a drain for leachate.

Black Soldier Fly: (*Hermetia illucens*) Generally unobtrusive or even beneficial to worm bins. These flies are very common. Their maggots have similar detoxifying properties to the worms. They also repel other pests.

Castings: The excrement of composting worms, and their egg cocoons. Castings are the main product of worm composting. They are rich in nutrients and fertilize soil.

Clitellum: The tan-colored band around the top center of the worm. Its presence is a sign of sexual maturity in worms. If a worm is severed past this point, it has a chance of surviving.

Cocoons: A part of the worm life cycle. Cocoons are the discarded clitellum of the worm that receives sperm during reproduction. The cocoon holds several eggs.

Detritivorous: The technical term for an organism that feeds primarily on waste and decaying organic matter. Epigeic worms and black soldier flies are examples of detritivores.

Driftworm: (*Lumbricus rubellus*) An endogenic worm, these worms are not good for composting. These worms also carry the name Red Wigglers. They are bigger than their composting namesakes. They are an invasive species and should not be brought to North America.

Endogenic worms: A type of worm that burrows in soil near the top. Their burrows are more shallow than anecic worms, but they still aren't as high up as composting worms.

Epigeic worms: A worm type suitable for worm composting. These worms don't burrow, instead, they remain in the topsoil and leaf layers, breaking down organic material and providing nutrients naturally.

European Nightcrawler: (*Eisenia hortensis*) Red worms, larger than red wigglers. These worms are good for composting and fish bait. They can also survive in saltwater, so they are very common for ocean fishing. There are approximately 400 worms per pound in weight.

Hermaphrodite: An organism that has the physical genitalia of both male and females. Most worms are hermaphrodites.

Indian Blue Worm: (Perionyx excavatus) A worm famous among vermicomposters for escaping bins, these worms are suitable for composting, but are very unpopular. If you have a better option, take it.

Lime Mix: An anti-acidic additive made of Calcium Carbonate. Very useful for overly acidic worm bins.

Microbes: Microscopic organisms. Bacteria, viruses, fungi, and protozoa are all examples of microbes. In composting, microbes are what break down food in order for worms to be able to eat it. They are also responsible for most decomposition processes such as fermentation and digestion.

Red Wiggler: (Eisenia fetida alt. foetida) The most common worm species used for composting. They are small and low cost, and are available almost everywhere in the world.

There are approximately 1,000 worms per pound.

Red Worm: A nickname that surprisingly describes several different species of worms. Be careful buying worms that are just known by this name, as it's a tossup which species you will actually receive. Remember to use Latin names.

Spigot: A tap or faucet to allow leachate drainage.

Stacked Worm Farm: A form of worm bin that consists of sections stacked one on top of the other. These are usually bought pre-made. The advantage of this system is that the worms are usually separated from the compost, for easy harvesting. These bins usually use the continuous flow method.

Tiger Worm: (*Eisenia andreia*) These worms are similar in appearance and colloquial name to red wigglers. They are good composting worms and are also similar in size to red wigglers. There are approximately 1,000 worms per pound.

Vermicompost: Compost made up of the castings of worms. Highly nutrient-rich and valuable, vermicompost is an excellent fertilizer and trade good. Also referred to as worm compost or vermicast.

Vermiculture: The practice of farming worms, harvesting their castings, and using the castings as fertilizers for various applications.

Worm Farming: The practice of breeding worms to be sold for a variety of purposes, such as composting or fishing bait. Vermiculture involves worm farming, but not all farming is vermiculture.

Worm Tea: A nutrient-rich liquid made by steeping vermicompost in water for long periods. The resulting liquid is a plant fertilizer. Also referred to as vermitea.

Conclusion

Congratulations! By now, you should be a vermicompost expert! You know how to start your first worm bins, the many benefits to doing so, how to troubleshoot any issues that come your way, how to make money vermicomposting, and even how to teach the craft to children. I'm so glad to have another vermicomposter for our wonderful community. If you're anything like me, and since you've read this whole book you probably are, you're itching to start a hobby that will certainly last you a lifetime.

Together, we've gone over the basics of what vermiculture is. You've learned how it works. You know what microbes are and how they help worms digest their food. You've also learned what materials you need to become a vermicompost master, as well as how to build and maintain your very own composting bin. You've learned a ton about worms: how they

GEOFF EVANS

work, how they eat and mate, and what makes their castings so special. You've also learned about several different species of worm and what makes them all unique. You know how to get rid of bugs in your bin that could hurt your worms. You have learned how to troubleshoot your problems when anything goes wrong. You have even gained some new entrepreneurial knowledge, so you can start your own vermicompost business. You know all the benefits of vermicomposting and how to share those benefits with friends and kids of any age. If you wanted, you could go and fulfill your dream of buying a ranch and some land and living out in the country. Just you, your family, and about ten thousand worms! You have officially earned the right to call yourself a Vermiculturist.

When I first started composting, I knew nothing about anything that I was doing. Over the years, I've learned so much. I've made mistakes and maybe killed a few too many worms, but now I know enough to pass on all those lessons I've learned. It warms my heart that my success can now be shared. I hope you take the expert knowledge I've shared with you and use it. Find contentment and joy within these pages. Share it with those you love.

It's been a wonderful journey to take together, and I'm happy you've joined me. You've received a truly comprehensive education on worm composting. The tips, tricks, and secrets contained in these pages are enough to guarantee success with any vermicomposting endeavor. If you've enjoyed reading this book as much as I enjoyed sharing my knowledge with you, **please leave a positive review on Amazon**. It helps me learn how to improve as an author, so I can teach even more curious souls the wonders of vermicompost.

Yours, Geoff Evans

Acknowledgements

I would like to acknowledge my wonderful friend Caroline, for helping me to edit and make this book legible. I would also like to thank my wife and children for being so utterly supportive of me throughout the creation of this non-fiction novella that I was so intent on giving to the worm composting community, and to the world. And finally, to you, the reader for getting all this way to the end. I hope you've gleamed value from it and are now ready to become a lover of this fantastic hobby too.

References

Addict, S. C. W., & Gardener, L. N. L. (2020, April 30). Harvesting Worm Castings: 4 Foolproof Methods. Epic Gardening. https://www.epicgardening.com/harvesting-worm-castings/

All Things Organic. (n.d.). Worm Bin Care and Maintenance. https://www.allthingsorganic.com/worm-bin-care-and-maintenance

Arnold, I. (2020, May 5). Packing Earthworms for Sale and Shipment. ToughNickel. https://toughnickel.com/self-employment/EARTHWORMS-PACKING-THEM-UP-FOR-SALE-AND-SHIPMENT

Bentley. (n.d.). Do Worm Bins Need to be Mixed? Red Worm Composting. https://www.redwormcomposting.com/reader-questions/do-worm-bins-need-to-be-mixed/

Bentley. (n.d.). VermiPonics. Red Worm Composting https://www.redwormcomposting.com/gardening/vermiponics/

Best Organic Fertilizer LLC. (2013). The Benefits Of Worm Compost. http://www.best-organic-fertilizer.com/the-benefits-of-worm-compost.html

Bhandari, S. (2018, May 20). What are gut bacteria? WebMD. https://www.webmd.com/digestive-disorders/qa/what-are-gut-bacteria

Bond, C. (2016, June 16). 3 Vermicomposting Fails & How To Do It Right. Hobby Farms. https://www.hobbyfarms.com/3-vermicomposting-fails-how-to-do-it-right/

Brown, P. (2020, January 11). Vermicomposting: How Many Worms Are Needed? Thriving Yard. https://thrivingyard.com/how-many-worms/

Casley, N. (2020, January 8). What is bokashi bran (compost accelerator)? Bokashi Living. https://bokashiliving.com/what-is-bokashi-bran/

Clatworthy, J., Hinds, J., & Camic, P. M. (2013, November 29). Gardening as a Mental Health Intervention: A Review. Mental Health Review Journal, 18(4), 214-225. doi:10.1108/mhrj-02-2013-0007

Cody. (2019, August 28). Worm Bin Maintenance & Upgrades. Thistle Down Farms. https://thistledownsfarm.com/worm-bin-maintenance-upgrades/

Donal O'Leary. (2019, November 20). How does a wormery work? Waste Down. https://wastedown.com/how-does-wormery-work/

Francesca. (2017, December 13). 5 Great Reasons To Split Your Worm Bin. The Squirm Firm. https://thesquirmfirm.com/5-great-reasons-to-split-your-worm-bin/

Francesca. (2017, June). 5 Simple Ways To Use Worm Compost In Your Garden. The Squirm Firm. https://thesquirmfirm.com/5-simple-ways-use-worm-compost-garden/

Fong, J., & Hewitt, P. (1996). More About Worms... And Related Classroom Activities. The Cornell Waste Management Institute. http://compost.css.cornell.edu/worms/moreworms.html

Fong, J., & Hewitt, P. (1996). Worm Composting Basics. The Cornell Waste Management Institute. http://compost.css.cornell.edu/worms/basics.html

Grant, B. L. (2018, April 5). Problems With Vermicomposting: How To Deal With Vermicompost Issues. Gardening Know How. https://www.gardeningknowhow.com/composting/vermicomposting/problems-with-vermicomposting.htm

Harrison, V. (n.d.). Worm Composting - The Basics Of Brewing & Using Worm Tea. Allotment & Gardens.

https://www.allotment-garden.org/composts-fertilisers/worm-composting/worm-composting-brewing-using-worm-tea/

Harroch, R. (2018, July 17). The Complete 35-Step Guide For Entrepreneurs Starting A Business. Forbes. https://www.forbes.com/sites/allbusiness/2018/07/15/35-step-guide-entrepreneurs-starting-a-business/

Hoskins, R. (2019, April 25). Do worms have eyes? And other worm facts. Woodland Trust. https://www.woodlandtrust.org.uk/blog/2019/04/do-worms-have-eyes/

Hundley, L. (2015, July 28). Poop? Or no poop? Compost Instruction https://www.compostinstructions.com/poop-or-no-poop/

Institute for Quality and Efficiency in Health Care (IQWiG). (2019, August 29). What are microbes? National Center for Biotechnology Information. https://www.ncbi.nlm.nih.gov/books/NBK279387/

Jackson, J. (2017, November 21). Are Worms Blind? Wild Sky Media https://animals.mom.me/worms-blind-11562.html

Jensen, J. (1997). K.I.S.S. Plan for Organic Farms, Dairies, or Other Large-Scale Operations. Happy D. Ranch.

http://www.happydranch.com/articles/Worm_Windro w_Method_For_High-Volume_Vermicomposting.htm

Jim, U. (2017, December 13). Foods That Can Hurt Composting Worms. Uncle Jim's Worm Farm. https://unclejimswormfarm.com/foods-hurt-composting-worms/

Jim, U. (2016, January 19). Vermicomposting Bin Maintenance for indoor and outdoor worm bins. Uncle Jim's Worm Farm. https://unclejimswormfarm.com/vermicomposting-bin-maintenance-for-indoor-and-outdoor-worm-bins/

Jim, U. (2016, February 25). Why are My Composting Worms Trying to Escape? Uncle Jim's Worm Farm. https://unclejimswormfarm.com/why-composting-worms-escape/

Johnson, S. (2017, March 28). Earthworms are more important than pandas (if you want to save the planet). The Conversation Canada. https://theconversation.com/earthworms-are-more-important-than-pandas-if-you-want-to-save-the-planet-74010

Journey North. (n.d.). Life of an Earthworm. University of Wisconsin-Madison Arboretum. https://journeynorth.org/tm/worm/WormLife.html

Kookaburra Worm Farm. Products: Garden Worm Bomb, Bag o Worms. Retrieved May 2020, from https://www.kookaburrawormfarms.com.au/products/

Michigan SoilWorks. Pricing. Retrieved May 2020, from https://michigansoilworks.com/pricing/

Napolitano, Jo. (2010, August 19). Exploring the Role of Gut Bacteria in Digestion: Better Knowledge of Tiny Stomach Stowaways Could Improve Human Health. Argonne National Laboratory. https://www.anl.gov/article/exploring-the-role-of-gut-bacteria-in-digestion

National Geographic Education Resource Library. (2012, October 9). Humus. https://www.nationalgeographic.org/encyclopedia/humus/

Nature's Footprint. (n.d.). Bugs! Creepy Crawlies! What's in my Worm Bin? https://naturesfootprint.com/community/articles/bugs-creepy-crawlies-whats-in-my-worm-bin/

NC Worm Farm (n.d.). FAQ. http://www.ncwormfarm.com/faq.html

Ophardt, C. E. (2003). pH Scale. In Virtual Chembook. Elmhurst College. http://chemistry.elmhurst.edu/vchembook/184ph.html

OrAgGrow Inc.(n.d.). Vermicomposting. OSC Organic Soil Corporation. http://oregonsoil.com/OrAgGrow/history.html

Owen, H. (n.d.). Feeding Your Worms. Worm Composting Hq. https://www.wormcompostinghq.com/feeding-your-worms/

Owen, H. (n.d.). Using Worm Compost. (n.d.). Worm Composting Hq. https://www.wormcompostinghq.com/how-to-use-worm-compost/

Owen, H. (2015, August 14). What is Worm Leachate? Is it Safe for Plants? Worm Composting Hq. https://www.wormcompostinghq.com/worm-bin/what-is-worm-leachate/

Owen, H. (n.d.). Worm Bin Bedding. Worm Composting Hq. https://www.wormcompostinghq.com/caring-for-worms/worm-bin-bedding/

Pauly. (n.d.). Leachate vs Worm Tea. Worm Farming Revealed. https://www.wormfarmingrevealed.com/leachate-vs-worm-tea.html

Pavlis, R. (2015, May 18). Compost accelerators, starters, activators, boosters. Garden Myths.

https://www.gardenmyths.com/compost-accelerators-starters-and-activators/

Pryce, C. (2019, October 11). How to create a wheelie bin worm farm. Wheelie Bin Solutions. https://wheeliebinsolutions.co.uk/blogs/news/how-to-create-a-wheelie-bin-worm-farm

Sherman, R. (2017, April 22). How do earthworms eat and poop - and other surprising facts. The Washington Post https://www.washingtonpost.com/national/health-science/how-do-night-crawlers-eat-and-poop--and-other-surprising-facts/2017/04/21/ebe2a8ac-13ee-11e7-ada0-1489b735b3a3_story.html

Science Learning Hub. (2012, June 12). Earthworms' role in the ecosystem. The University of Waikato Te Whare Wānanga o Waikato https://www.sciencelearn.org.nz/resources/9-earthworms-role-in-the-ecosystem

Smith, M. R. (2019, June 21). How to make your own wormery. Vertical Veg https://verticalveg.org.uk/how-to-make-your-own-wormery/

Steve. (2016, September 22). How Do I Know If My Vermicompost is Ready to Harvest? Urban Worm Company. https://urbanwormcompany.com/vermicompost-finished-ready-harvest/

Steve. (2019, April 12). Vermicomposting 101: Should I Mix Worm Species in My Worm Bin? Urban Worm Company. https://urbanwormcompany.com/should-i-mix-worm-species-in-my-worm-bin/

Steve. (2019, July 1).Vermicomposting: The Ultimate Guide for the Beginner and Beyond. Urban Worm Company. https://urbanwormcompany.com/vermicomposting-ultimate-guide-beginner-expert/

Soil Association. (n.d.). Why are worms important? https://www.soilassociation.org/our-campaigns/save-our-soil/meet-the-unsung-heroes-looking-after-our-soil/why-are-worms-important/

Sustainable Jungle. (2020, January 12). How To Make Worm Tea: The All-Natural Boost Juice For Plants. https://www.sustainablejungle.com/sustainable-living/worm-tea/

The Squirm Firm. (n.d.). Worm Composting Accessories. https://shop.thesquirmfirm.com/worm-composting-accessories/

The Tasteful Garden. (n.d.). Worm Castings - What are they? How do they work? https://www.tastefulgarden.com/Worm-Castings-d114.htm

Tucker, J. (2019, December 03). LLC vs. Sole Proprietor: How to Make the Right Choice. Nav.com. https://www.nav.com/blog/llc-sole-proprietor-18376/

U of I Extension. (n.d.). Can't Live Without Me: The Adventures of Herman the Worm. https://web.extension.illinois.edu/worms/live/

Uncle Jim's Worm Farm. Worm Hobby Kit. Retrieved April 8, 2020, from https://unclejimswormfarm.com/product/composting-worms/worm-hobby-kit/

United States Environmental Protection Agency. (2019, December 18). Basic Information: Pathogen Equivalency Committee. https://www.epa.gov/biosolids/basic-information-pathogen-equivalency-committee

Vanderlinden, C. (2019, December 8). Why Are My Worms Trying to Escape From My Vermicompost Bin? The Spruce https://www.thespruce.com/worms-escaping-from-my-vermicompost-bin-2539483

Vinje, E. (n.d.). Worm Composting: A Beginner's Guide. Planet Natural Research Center. https://www.planetnatural.com/worm-composting/

Working Worms. (2016). Money from Farming Worms http://working-worms.com/money-from-farming-worms/

Working Worms. (n.d.). What About the Workers? - Earthworm Versus the Redworm. http://working-worms.com/what-about-the-workers-earthworm-versus-the-redworm/

Working Worms (n.d.). Worm Reproduction http://working-worms.com/worm-reproduction/

Working Worms (2016). Vermiculture. In The Worm Dictionary and Vermiculture Reference Center. http://working-worms.com/the-worm-dictionary-and-vermiculture-reference-center/

Worm Composting Canada. Shipped Order Pricing. Retrieved May 2020, from https://www.wormcomposting.ca/composting-worms/shipped-order-pricing/

Wormcity. Wormery Frequently Asked Questions. Retrieved May 2020, from https://www.wormcity.co.uk/wormfaq.htm

Wormery. (n.d.). Customer Service FAQs. https://www.wormery.co.uk/wormery-faqs

CPSIA information can be obtained
at www.ICGtesting.com
Printed in the USA
LVHW031225110121
676185LV00003B/204

9 781913 666088